BASIC METHODS IN

Molecular Biology

BASIC METHODS IN

Molecular Biology

Leonard G. Davis, Ph.D.
Neurobiology Group
Central Research and Development Department
E.I. Du Pont Experimental Station
Wilmington, Delaware

Mark D. Dibner, Ph.D.
Director, Business Studies Program
North Carolina Biotechnology Center
Research Triangle Park, North Carolina

James F. Battey, M.D., Ph.D.
National Cancer Institute
Medical Oncology Branch
National Naval Medical Center
Bethesda, Maryland

APPLETON & LANGE
Norwalk, Connecticut

Cover Illustrations:

Representations of 24 base pairs of the standard "B" form of DNA, photographed on an Evans and Sutherland PS300 (Arnott, S., and Hukins, D., Biochem. Biophys. Res. Comm. *47:*1504, 1972). The molecular surface is displayed with dots (Connolly, M. L., Science *221:*709, 1983). Color coding is by atom type: nitrogen is blue; carbon is green; oxygen is red; phosphorus is yellow. The back cover shows the same molecule, cross-sectioned approximately halfway through the helix. Cover illustrations were created by and are courtesy of Dr. J. M. Blaney of the Biomedical Products Department of E. I. du Pont de Nemours and Company, Wilmington, Delaware.

93 94 95 96 97 / 10 9 8 7 6 5 4 3 2 1

Prentice Hall International (UK) Limited, *London*
Prentice Hall of Australia Pty. Limited, *Sydney*
Prentice Hall Canada, Inc., *Toronto*
Prentice Hall Hispanoamericana, S.A., *Mexico*
Prentice Hall of India Private Limited, *New Delhi*
Prentice Hall of Japan, Inc., *Tokyo*
Simon & Schuster Asia Pte. Ltd., *Singapore*
Editora Prentice Hall do Brasil Ltda., *Rio de Janeiro*
Prentice Hall, *Englewood Cliffs, New Jersey*

ISBN 0-8385-0582-1

Library of Congress Catalog Card Number: 92-056591

PRINTED IN THE UNITED STATES OF AMERICA

Contents

Foreword

The heart of the most recent revolution in biology has been the development of the technology of genetics. Its achievements have simply changed what biologists do and, perhaps even more important, the way they think. Moreover, never before have scientists from such a broad range of disciplines rushed into such a small and slightly arcane field (as molecular geneticists used to believe theirs was) to learn, to carry off a bit of the technology, and to do it quickly because, armed with these powerful tools there was so much to do, so much to be learned. Doctors Davis, Dibner and Battey have done us a great service in providing the most powerful tool of all—an up-to-date, accessible, laboratory-tested, and comprehensive embodiment of what one needs to know to get on with the job at hand. They are experienced scientists. They state the principles and give the details. The rest is up to us.

Philip Leder
Boston

Acknowledgments

The authors wish to thank the many people and institutions that assisted us in compiling this text.

A number of scientists made valuable contributions to this text and deserve due recognition. Dr. Shoshana Segal provided the sections on mammalian cell transfection. Ms. Marian Kelley was the source of technical detail on monoclonal antibody preparation. Dr. Hunt Potter provided valuable help toward our writing of the section on electroporation. Ms. Donna Reed was a valuable resource for the preparation of protein methods. Dr. Eric Sinn contributed a large portion of our section on transgenic mouse analysis, and Dr. S. Carl Falco provided information for our comments on the use of yeast. Dr. Edward Sausville assisted in the preparation of the sections on cDNA cloning. The cover illustration was provided by Dr. J. Blaney.

Many other people provided assistance in the review of our manuscript. They include Drs. Philip Leder, Marion Cohen, Edward Berger, Rick Woychik, Keith Lawrence, Eric Seifter, Anne-Marie Lebacq, and Michael Kuehl. Sections of this manual were reviewed by Drs. J. Angulo, R. Arentzen, M. Lewis, F. Baldino, Jr., and L. Hennighausen. Additional help in reviewing was provided by A. Callahan, R. Manning, and R. Lampe.

We thank the people of Elsevier Science Publishing, and especially Mr. Yale Altman and Mr. Jonathan Wiener for their help and support in compiling this, our first book. Mr. Brian Trench created many of the drawings herein. We also thank New England Bio Labs, Bethesda Research Labs, Bio-Rad, Boehringer Mannheim, and Pharmacia for permission to reproduce selected illustrations. We thank Ms. Sherry M. Vari for her excellent word processing skills and long hours spent on text preparation.

Most of all, the authors are indebted to Dr. Philip Leder, Chairman of the Department of Genetics of the Harvard Medical School, for providing the leadership and resources responsible for the generation and evolution of most of these methods. We also thank past and present members of his laboratory group for sharing their knowledge with us to make this book possible.

Lastly, we wish to dedicate this book to our wives, Penny, Elaine, and Fran, who were wonderfully supportive of this project and the time it required.

Leonard G. Davis
Mark D. Dibner
James F. Battey
—April 1986

SECTION

The Basics of Molecular Biology

The Basics of Molecular Biology

Current biological science has been revolutionized by a series of new investigative techniques developed within the last 15 years. These techniques allow the definition of molecular mechanisms and structures that are responsible for such complex processes as cell growth and division, metabolism, differentiation and development. More significantly, they provide a way to manipulate molecules critical to these processes, and observe the changes in living systems that incorporate the altered molecules.

Nucleic acids and proteins are macromolecules; linear polymers comprised of subunits. Nucleic acids encode the genetic information specifying the primary structure of all proteins unique to an organism. Together with lipids and extracellular supporting stroma, they create cellular activity and physiological function. Thus, biological functions can be understood in part by examining the interrelationships between these key components. The genetic material of the cell, *deoxyribonucleic acid* (DNA), is a polymer composed of four nucleotide building blocks. Each of the four nucleotides contains a nucleic acid base (A, adenine; G, guanine; T, thymine; C, cytosine), a deoxyribose sugar moiety, and a phosphoester. Each strand is a string of nucleotides covalently bound together by phosphoester linkages between the 5′ carbon on the deoxyribose sugar of one nucleotide and the 3′ carbon of the sugar moiety on the neighboring nucleotide. Chains of these DNA subunits exist as two antiparallel strands in opposite polarity with respect to the phosphate sugar backbone, wound around each other in a double helical structure. One strand binds tightly to the other strand because there is the potential for hydrogen bond formation between specific bases on one strand with bases on the opposite, or complementary, strand. Adenine is always paired with thymine, and guanine with cytosine. The fidelity of base pairing is provided by the nucleic acid synthesizing machinery that normally adds only the "correct" base specified by the template strand when elongating a new strand. It is the constancy and specificity of this complementary base pairing that forms the basis of DNA's function as a repository of genetic information. The order of nucleotides in DNA corresponds to the order of amino acids in proteins. As such, DNA can encode for proteins, with triplet groups of three adjacent nucleotides representing an mRNA codon, which specifies a particular amino acid. Therefore, the linear nucleotide sequence in DNA specifies the order of amino acids for the cell's structural, functional, and enzymatic proteins. Other regions of DNA, which do not directly encode protein, contain information directing the regulation of gene product synthesis.

In the synthetic pathway between DNA and protein are the ribonucleic acids (RNA). The strand encoding the protein sequence information of the double-stranded DNA is copied, or transcribed, into a complementary strand of RNA. This RNA contains the same bases as DNA, except that uridine (U) is substituted for T and a ribose moiety is present instead of the deoxyribose. The RNA copy of the gene, called messenger RNA (mRNA), is translated with the assistance of transfer RNA (tRNA) and ribosomes (rRNA and associated proteins) to assemble sequentially the amino acids that form the primary sequence of protein.

Many molecular biology laboratory methods take advantage of the relative simplicity of prokaryotic cell systems such as bacteria. In prokaryotes, the continuous linear DNA sequence corresponds directly to linear RNA and protein sequences. However, in eukaryotes, the DNA encoding for protein cannot be read continuously as it contains interruptions (introns) in the translatable sequence. Eukaryotic DNA is thus first copied to a primary transcript (heteronuclear RNA) that is processed in the nucleus by excision of the protein coding sequences (exons). The exons are joined linearly into mature mRNA that can be processed further in the nucleus and moved to the cytoplasm for translation into protein. Certain newer methods allow the study of genes in eukaryotic cell systems.

Understanding the structure, function, and regulation of genes and their products is essential to an appreciation of biological systems. This also involves understanding the organization of an organism's nucleic acids. Previously this understanding was confounded by the complexity of the genome in eukaryotic cells, which contains up to 10^9 nucleotides in 50,000 genes. To analyze the genetic structure and events in this complex situation, one needs the ability to isolate and study a single gene in a purified form. Molecular cloning of DNA provides a mechanism for isolating a single discrete segment of DNA from a population of genes, purifying this segment to homogeneity, and amplifying the DNA segment to produce enough pure material for chemical, genetic, and biological analysis. The process of cloning relies entirely on performing enzymatic reactions in the laboratory, using well characterized bacterial DNA cleaving enzymes (restriction enzymes, REs) and modifying enzymes to copy, cut, an[illegible] splice together discrete DNA molecules. DNA molecules are thus introdu[illegible] into bacterial cells after being spliced into autonomously replicating DN[illegible] cles (plasmids) or bacterial viruses (bacteriophages). After many rou[illegible] replication, the hybrid molecules are reisolated and purified, yielding [illegible] quantities of the cloned DNA segment.

With the isolated, purified DNA segment the nucleotide seque[illegible] can rapidly be determined, leading to the prediction of the amin[illegible] of the encoded protein. Radioactive labeling of this purifie[illegible] scientist to specifically probe for copies of related DNA se[illegible] cell genomes or related intracellular mRNA, amidst a b[illegible] million unrelated sequences. mRNA synthesis from t[illegible] detected and quantitated in amounts as low as o[illegible]

Reengineering of the cloned DNA in bacteria or yeast may allow expression of its protein coding sequence, providing an inexpensive and abundant source of otherwise unattainable proteins of biological or medical importance. Alternative versions of the cloned DNA can be created in the laboratory by changing the structure or sequence. These DNA constructs can then be reintroduced into cells or whole animals to study the results of these man-made changes or mutations, and understand more completely the function and regulation of genes.

In this book, we describe methods for performing these experiments in molecular genetics. In each case, the method is described in a step-by-step, "cookbook" format and has been used, as written, with favorable results.

A WALK THROUGH THIS MANUAL

The methods in this book range from very simple to very complex. First is a description of the plasmid and vector systems and bacterial host cells used in the methods. The initial sections assume that a specific synthetic or cloned DNA probe is already available, allowing the selection, amplification, and examination of the gene of interest. Methods for isolating DNA from tissue, cutting the DNA to usable size, and separating the DNA pieces by size are discussed in Section 5. Sections 6 and 7 present methods for making probes, either synthetic or plasmid derived, to use in selecting DNA of interest. Methods for plasmid preparation and amplification are presented in Section 8. From the amplified plasmids, cloned DNA is excised and purified (Sections 9 and 10).

Section 11 turns to RNA—its preparation, selection, separation, and analysis. In Section 12, another type of cloning vector, the bacteriophage, is described. Please note that up to this point, the methods described involve the selection and amplification of DNA sequences that have already been cloned. The next two sections, 13 and 14, present methods for creating genomic DNA and cDNA libraries in bacteriophage vectors.

From the created library a desired clone is selected. The next step is to grow that DNA on a large scale, as described in Section 15 on subcloning into plasmids for preparative growth. From the higher yield of this cloned DNA, the sequence and other properties can be studied, following cloning into an appropriate M13 vector (Sections 16 and 17). Up to this point, DNA has been studied using the benefits of simpler prokaryotic systems. However, it may be of interest to put modified versions of the cloned gene back into the genome of eukaryotic ·ells in order to evaluate its regulation and function in a more biologically ·vant system. Section 18 describes methods for incorporating DNA into mam- ·ells growing in culture.

·ntioned above, proteins are the product of the genetic material, and it ·nt to study them in order to understand gene regulation. Also, it ·ate RNA into proteins in vitro. These protein-related meth- ·tion 19.

·methods (20) describes basic techniques that are ·er methods discussed in the text, such as DNA titration of plaques. It is anticipated that the novice will refer to these methods initially; in time they will become second nature.

Lastly, several more specialized molecular biological methods are described in Section 21. The first, transgenic mouse analysis, involves incorporation of new DNA pieces into a mouse embryo for later analysis in the postpartum animal. We also describe monoclonal antibody production techniques used to prepare immunological probes for specific gene products, as well as in situ hybridization, which uses nucleotide probes to localize and study specific genetic messages in tissue sections. Finally, some general notes are given on the use of yeast host and vector systems to perform molecular biology techniques.

The next few pages describe the use of specific techniques in molecular biological studies, with attention to questions that can be addressed using these methods.

SECTION

2

The Tools of the Molecular Biologist

The Tools of the Molecular Biologist

To illustrate the use of molecular biology methods, this section follows one possible series of experiments to study a typical gene, *X*, employing a variety of these methods.

It may be desirable to study gene *X* for its interesting structure or relevant expression in some biological context. Initially, a radiolabeled DNA probe needs to be obtained with a sequence similar to that on gene *X*, for example the gene from another species (homology to gene *X*). This probe can be purified and nick-translated to form a radiolabeled probe in order to detect the presence of gene *X* in a Southern blot analysis. Alternatively, a synthetic oligonucleotide probe can be synthesized in the laboratory to contain a sequence complementary to a portion of gene *X*. The labeled probe can then be used in DNA blotting to analyze DNA from a tissue or cell line of choice using DNA blots to define the presence of gene *X*–related sequences in the genome.

To do these DNA (Southern) blots, DNA from a tissue or cell line is isolated and purified and cut with specific restriction endonuclease(s) (REs) into defined fragments; the fragments of DNA are then fractionated by size using agarose gel electrophoresis. The DNA on the gel is transferred to a nitrocellulose filter (Southern blot), and the blot is hybridized with probe specific for gene *X* (Southern hybridization). The probe forms complementary base pairs only with restriction fragments that contain homologous sequences. Nonspecific radioactivity is washed away, and autoradiography of the blot demonstrates one or more bands if gene *X* is present or no bands if gene *X* is not found in the DNA tested.

An altered pattern of hybridizing DNA restriction fragments may appear on the Southern blot from DNA made from a specific tissue sample, indicating a change in the gene *X* structural sequences. For example, if there is a rearrangement of DNA in a specific tissue or tumor, this "somatic" rearrangement can be identified by purifying DNA from different tissue sources and probing, as described above. Genomic DNA from different cell types or tissues might show different size hybridizing fragments on the Southern blot, resulting from the changes introduced by rearrangement in the DNA.

Another example of an altered DNA pattern might be due to restriction fragment length polymorphisms (RFLPs) or different gene forms (alleles). If the genomic DNA from 100 individuals was cut with the RE *Eco*RI and was probed on Southern blots with the probe for gene *X*, two or more different subpopulations may appear. Therefore, there may be some individuals whose genomic DNA contains a 10-kb hybridizing restriction fragment, others with a 6-kb hybridizing fragment, and heterozygous individuals with both 6- and 10-kb fragments, representing the two different alleles for the locus of gene *X* present in the population studied. Some of these RFLPs may be linked to genes responsible for genetically based diseases or a predisposition to malignancy. A correlation between detection of a specific allele of a gene *X* and developing a genetic disease or malignancy could be established. Southern blotting studies then allow an estimation of the likelihood of developing the disease. Genetic determination of RFLPs linked to Huntington's disease, Duchenne's muscular dystrophy, and cystic fibrosis has recently been demonstrated, allowing identification of individuals who are likely to develop these diseases before they reach reproductive age, or even before birth.

To understand more about the detailed structure of gene *X* or its altered structure under different biological conditions, one would need larger quantities of a homogeneous preparation of DNA from gene *X*. A genomic library is created by generating a random collection of fragments that represent all regions of the genome at least once. These fragments are individually inserted into an autonomously replicating prokaryotic DNA species, or vector, such as a bacteriophage λ derivative. The recombinant library members are used to infect bacterial cells. These cells divide many times, with each cell containing multiple copies of the insert-bearing phage. In most instances, at least a few of the cloned fragments will contain gene *X*. The same radiolabeled probe as above can be used to isolate clones for gene *X* from a genomic library.

A bacteriophage library of cloned genomic fragments is thus generated, with each of the million or so bacteriophage in the library producing multiple copies of its unique inserted DNA fragment. To figure out which of these phage in the library is replicating gene *X*, the entire library must be screened. This is done by plating the entire library on culture dishes, transferring a small portion of the DNA produced in each bacteriophage plaque to a nitrocellulose filter, hybridizing the gene *X*–specific probe to the filters, and autoradiographically identifying the hybridized DNA as darkened spots on the X-ray film. Any spots detected on the autoradiographs identify bacteriophage clones of gene *X*, or *X*-related sequences, and allows for isolation and further purification. After a few more rounds of screening, it is possible to pick out homogeneous clones for gene *X*. The sample now contains only one type of bacteriophage, the one with gene *X* inserted. The gene *X*–bearing fragment can be excised from the bacteriophage genome using an RE, and this fragment can now be reinserted into a more compact vector, such as a plasmid. With the DNA of interest inserted in the plasmid, one can now propagate this cloned DNA in a preparative fashion to make milligram quantities of gene *X*, to facilitate further study.

Using the genomic DNA clone, it is possible to look at structural landmarks and determine the nucleotide sequence of gene *X*. For example, samples of gene *X* can be cut by different REs to map the positions of RE sites within and around gene *X* (RE cleavage sites are highly specific for a given short DNA sequence). To do this, gene *X* DNA is digested by a combination of REs and the digested samples are electrophoresed on an agarose gel. Comparison of the sizes of fragments generated by multiple RE digests performed together or separately will yield an RE site map of the genomic DNA clone.

Using the RE cleavage map generated for gene *X*, small DNA fragments from *X* can be prepared and cloned into M13 bacteriophage vectors. These fragments can be sequenced using either the dideoxy chain termination or chemical degradation technique. From the sequences, it is possible to gain information about coding regions (exons) and noncoding regions (introns) on gene *X* as well as to identify known regulatory elements. Comparing the sequences to the universal genetic code allows the determination of the amino acid sequence encoded within gene *X*.

The same labeled probe for gene *X* can be used to look at RNA from eukaryotic cells of interest in order to understand better how gene *X* is expressed. RNA can be isolated and purified from cells and mRNA, whose sequence contains a string of A bases on its 3′ end (poly A tailed), can be purified from rRNA and tRNA based on its specific binding to an oligo-dT column. This mRNA may then be size fractionated on a denaturing agarose gel, transferred to nitrocellulose by blotting (Northern blot), and a probe homologous to gene *X* can be hybridized to the blot. These experiments demonstrate both the size of gene *X* transcripts and the relative amounts of the expression of gene *X* between tissues. Using this method, one can follow the expression of gene *X* over time, between tissues, or the regulation by an inducing or repressing agent. Moreover, the size of mature mRNA or an mRNA precursor can be determined for the products of gene *X* by Northern blot techniques.

If the protein product of gene *X* is characterized, additional methods can be employed to study the gene. An antibody to the gene product can be made for use as a probe. Thus, to affirm that the mRNA isolated from gene *X* is an appropriate substrate for translation, the mRNA can be translated in vitro to its protein product, and the protein can be analyzed with an appropriate monoclonal or polyclonal antibody, or by using a functional biological test. Using methods conceptually similar to DNA and RNA blotting, protein products can be fractionated by size on polyacrylamide gels and transferred to nitrocellulose for identification with an antibody (Western blot analysis).

Another more detailed analysis of RNA transcribed from gene *X* is provided by S_1 nuclease protection mapping. In this method, mRNA is hybridized to a ^{32}P-labeled, single-stranded DNA probe specific for gene *X*. These RNA-DNA (heteroduplex) structures are digested with the single-stranded DNA-specific endonuclease S_1, which digests all ^{32}P-labeled single-stranded DNA regions not protected by base pairing to RNA. The S_1-protected DNA regions are then resolved and sized on a denaturing gel, and then visualized by autoradiography. By using segments of the genomic clone for *X* as probes to hybridize to transcribed mRNA, the intron and exon regions can be more precisely determined. This analysis also gives information about the genomic location of the 5′ and 3′ end regions of gene *X*.

A definitive structural analysis of RNA transcripts derived from gene *X* is possible by obtaining a collection of complementary DNA (cDNA) clones made from mRNA in cells or tissue expressing this gene. These clones are usually isolated from cDNA libraries of clones that contain a representative sample of mRNAs obtained from a given cell or tissue. To generate a cDNA library from cell or tissue mRNA, the RNA-dependent DNA polymerase AMV reverse transcriptase is used to reverse transcribe a cDNA copy of mRNA. DNA duplexes are then formed using the enzymes RNase H and DNA polymerase I. This double-stranded DNA is then inserted into appropriate bacteriophage cDNA cloning vectors (λgt10 or λgt11). Each recombinant bacteriophage contains a cDNA copy from a single mRNA molecule. cDNA clones from gene *X* can then be isolated by hybridization to the *X*-specific probe using a method very similar to that employed for genomic clones. All structural changes and coding region boundaries in the transcribed *X* gene can be defined precisely by comparing the restriction map and nucleotide sequence (obtained as above) from the cDNA clones with the sequence of the genomic alleles of gene *X*.

The same cDNA libraries can be used to obtain the progenitor clone for a gene. These original clones may be selected by making a synthetic oligonucleotide hybridization probe with a nucleotide sequence inferred from a region of the encoded protein if a portion of the amino acid sequence is known. Another method is to construct a cDNA library in a protein fusion vector system, such as λgt11. In this cloning vector, the double-stranded cDNA is inserted near the carboxy end of the gene for the enzyme β-galactosidase. If the cDNA sequence is inserted in the appropriate translational reading frame and in the correct orientation (one time in six), this recombinant bacteriophage will synthesize a β-galactosidase fusion protein whose carboxy terminus will be the amino acid sequence specified by the cloned cDNA insert. cDNA clones made this way may be screened using antisera specific for the protein of interest. Clones making the fusion protein may react positively with antisera, allowing selection and purification to homogeneity.

Now that the structural and expression properties of gene *X* have been determined, this gene may be reinserted into mammalian genomes using DNA transformation techniques. Experiments of this type can yield the definition and analysis of regulatory elements in sequences flanking the gene. The insertion of modified versions of gene *X* into cultured cells can further test the function of these regulatory sequences. In addition to examining the regulation of gene *X* by studying its products, it is possible to examine the phenotypic changes in the host cell caused by insertion and altered expression of the gene.

Transgenic mouse analysis is an elegant and powerful extension of this methodology. With this procedure, DNA from gene *X* is microinjected into the fertilized egg of a mouse. Embryos with inserted DNA are implanted into host pseudopregnant mothers, and the offspring are characterized after birth. Genomic DNA from these transgenic mice can be examined for the stable integration of gene *X* in their germ line. In the mice that have this transgene, the regulation of gene expression and the effects of that gene on the host animal can be studied. A single gene can be studied in a variety of host cells (i.e., different tissues) in a manner that is extremely difficult to duplicate using cell culture. In addition, tissue-specific and developmental stage-specific gene regulatory elements can be identified, and their properties can be explored in the whole animal.

The techniques described in this manual provide the basis for the systematic study of mammalian gene expression. These methods are not always straightforward or simple to use, but once they are mastered, they will be powerful tools to help the researcher understand the functioning of biological systems. They range from simple isolation of DNA to actual placement of foreign DNA into the genome of a host animal. Correspondingly, they range from simple laboratory tools to those that are highly complex and still under development. Using these methods, it is possible to define the genetic basis of certain diseases, physiological reactions, biochemical mechanisms, and cellular processes. These tools are also being used for the generation of products of biotechnology, including powerful new therapeutics, new diagnostics, and other valuable biomedical products.

SECTION

3

General Preparations, Procedures, and Considerations for Use of Manual

SECTION

3-1. Using This Manual

The users of this manual will include scientists already engaged in research involving molecular biology, as well as students or scientists using these methods for the first time. For those just beginning, this section contains information to help them get started. For the experts, it is helpful to read the next few pages first in order to become familiar with items specific to this book.

This manual describes standard techniques that have been shown to work in the laboratory, as well as newer ones that are beginning to prove important. In order for this book to be kept at a reasonable length, certain information was placed in centralized locations. Thus, although the methods were designed to be complete, you may need to derive information from other parts of this book when attempting to complete a technique. For example, over half of the methods require extraction and precipitation of DNA or RNA. We have included a generic method on SS-phenol/chloroform extraction of DNA or RNA samples followed by ethanol precipitation. This submethod may not be given in detail within each method; refer to it specifically (Section 20-1), if necessary.

In addition, not every buffer has its formula within the specific method. Note that the "Reagents" section for each method contains some buffers or solutions whose names are in *italic* or in **bold** type. Those buffers appearing in *italic type* are stock solutions and their formulae appear in Appendix I. Buffers appearing in **bold type** are a dilution of a stock solution. Note also that ethanol refers to 95% or absolute ethanol, H_2O refers to distilled/deionized water, and, unless otherwise noted, all centrifugations are at room temperature. Any variations on this schema will be noted in the text. It is suggested that, before attempting to use any technique, the "Reagents" section be studied, with all reagents prepared in advance.

Suppliers of certain reagents and equipment are noted in parentheses in various places in the manual and listed in Appendix III. With the increasing number of suppliers, large and small, it is impossible to list all possible sources of each item. In general, the suppliers listed are potential sources. If more than one supplier is listed, the first is the one we use and find to work well. Keep in mind that this is not an endorsement, and the user must determine which source is best for any particular item. Also, quality and service may vary over time. The scientist must weigh many variables, such as reliability, service, quality, price, location, and responsiveness, in choosing a source for a specific item.

All methods have been written in stepwise form, as successfully used in our laboratories. Most methods have infinite variations, and the method, as shown, is only one of them—one proven to work well for us. Those who have tried some of these methods know that troubleshooting is often required to get a method to work optimally. Wherever possible, we have included hints for troubleshooting or for improving a procedure when something has gone wrong. Because it is impossible to replicate all aspects of a method (e.g., the same equipment, suppliers, batches of reagents, water supply), we have also tried to point out potential pitfalls or highly important aspects that should not be varied. These methods work for us, and it is hoped that, with little or no modification, they will work for you.

A major source of problems when attempting to start these procedures can be contamination, in solutions or otherwise. The best ways to prevent it are two: make sure you have a source of extremely pure water (and use this water in all steps), and wear surgical or laboratory gloves throughout every procedure.

References are included with the individual methods. In most cases, they are the original references and were a starting point for the development of the method described. In many instances, the method described evolved in many stages, and no reference may be available for the current stage.

Most methods include footnotes that contain important additional information. It is useful to read an entire method, including footnotes, from start to finish before beginning a procedure.

Lastly, note that in some ways these are generic methods. When a specific tissue or cell is called for, other tissues or cells may sometimes be substituted with only minor modifications. Once the techniques are established in the laboratory, they can be expanded or modified.

GENERAL METHODS USED IN MANY OTHER TECHNIQUES

A number of methods are used repeatedly as part of more complex techniques. Rather than describe each of them in detail every time they are encountered, we have written them as separate generic methods. Within the larger methods, the generic methods are referred to (e.g., "run a sample on minigel, as described in Section 9-1" or "electroelute DNA from the gel piece, as described in Section 9-3").

The following methods are often used and described separately:

Restriction endonuclease digestion (Section 5-4)
Southern blotting (Section 5-6)
Nick translation (Section 7-1)
DNA hybridization (Section 7-2)
Transformation of bacteria (Section 8-1)
Minigels (Section 9-1)
Electroelution (Section 9-3)

SS-phenol extraction/ethanol precipitation (Section 20-1)
Optical density measurement (Section 20-3)
Photographing gels (Section 20-4)
Autoradiography (Section 20-5)
Making plates for bacterial growth (Section 20-6)
Titering and plating of plaques (Section 20-7)

We suggest that these methods be learned or practiced before more complex methods are attempted. Regardless, it is likely that in very little time you will gain familiarity with these often used methods and will no longer have to refer back to their description when you encounter them in more complex methods.

ABBREVIATIONS AND JARGON TO UNDERSTAND WHEN USING THIS MANUAL

Only a few abbreviations are used in this manual. Other than those given below, the terms used are standard in most laboratories. For a full glossary of molecular biology terms, consult a molecular biology textbook.

A, T, G, C—the nucleotide bases adenine, thymine, guanine, and cytosine
bp–base pairs
blot—a transfer of DNA, RNA, or protein, generally from gel to a solid support, such as a nitrocellulose filter
dNTP—deoxynucleotide triphosphate (e.g., dATP, dCTP)
ddNTP—dideoxynucleotide triphosphate
ethanol—absolute or 95% ethanol
H_2O—pure (deionized/distilled) water
insert—a foreign sequence of DNA added to a phage or plasmid vector
kb—kilobases (1,000 nucleotide bases)
NC—nitrocellulose
N°—nucleotide mix with limiting amounts of a specific base (e.g., G° is a mix with less G bases)
O.D.—optical density read at a defined wavelength (e.g., 600 nm)
oligonucleotide—a short (e.g., <50-base) DNA sequence
pfu—plaque-forming units
RE—restriction endonuclease
U—unit of enzyme activity
UV—ultraviolet light
vector—an autonomously replicating cloning vehicle

Abbreviations for company names, such as BM, BRL, NEBL, P-L and others are defined in Appendix III.

The next few pages will cover tips on safety, and equipment needed.

SECTION **3-2.**

Safety Considerations

Safety is of primary importance in the laboratory, and these techniques can often be hazardous. Although certain dangerous aspects of the methods will be pointed out, this book cannot be considered a safety manual; the individual must understand all safety considerations and carry out laboratory operations with maximum safety. A full discussion of the safety considerations that are necessary for handling chemicals, radioactivity, and biological agents is beyond the scope of this book. Here are a few specific things to keep in mind:

Toxic chemicals. A number of chemicals used in these procedures are toxic and require special care. Many solvents, such as chloroform, isoamyl alcohol, isobutanol, *n*-butanol, formaldehyde, and ether, should be handled in a fume hood and with protection for clothing, skin, and respiration. Other reagents are potential carcinogens or mutagens, such as ethidium bromide or formamide, and require protection when handling. In addition, because of the toxicity of these and other agents, pipetting by mouth should never occur in the laboratory. Information concerning the safety of individual reagents should be available from the supplier, and all warnings should be heeded.

Radioactivity. A number of procedures require the use of ^{32}P or other radioactive substances. Make sure that all rules for the safe handling of such materials are strictly followed. Take equal precautions with radioactive waste. Your institution has rules governing the use of radioactive substances. It is advised that you be familiar with these rules as well as local procedures for the handling of radioactivity, including spills, should they occur. Make sure that proper shielding, storage, and disposal materials are available. In general, keep track of the location of all radioactivity at all times during a procedure and handle it accordingly.

Biological containment. Some organisms you may wish to work with may require a higher level of containment than the laboratory bench. You must be aware of these requirements and provide a higher level of containment. The National Institutes of Health Guidelines on Recombinant DNA Research provides specific recommendations for particular experiments. Be aware of any local or governmental ordinances regulating biotechnological processes. Know how to decontaminate and sterilize an area if a spill occurs.

Equipment hazards. Some of the equipment used in these techniques may be dangerous if not handled properly. Especially dangerous is the electrical shock from power supplies for electrophoresis. Again, consult the literature accompanying equipment for safety information.

For all of the above, information is available regarding potential hazards, toxicity, and safe handling from the suppliers of reagents, radioactive materials, and equipment. In addition, the Occupational Safety and Health Administration (OSHA) and your institution have information regarding potential hazards in the handling of most reagents. The researcher must be aware of all potential hazards and know how to deal with them safely.

SECTION **3-3.**

Equipment Needed for Molecular Biology Studies

There is a host of equipment required for the methods described in this book. Fortunately, most of the equipment is common to many biochemical or cell biology laboratories. Other items have fairly specific uses and are not generally found in a nonmolecular biology laboratory.

It would be impossible to list all of the items used for experimentation in molecular biology. The following list contains items considered most important for use.

IMPORTANT COMMON ITEMS

Incubators. Almost every method has multiple incubations. It is important to be able to switch from one temperature to another. Thus, at least two heating water baths are needed. Generally, a refrigerated bath is not needed; an ice bucket bath adjusted to the proper temperature will serve well. In addition, at least one dry incubator (oven) and one vacuum oven are required. Also, some incubations are done with shaking, rotating, swirling, and so on. Separate incubators or attachments for incubators that perform these functions are needed.

Cell growth equipment. Orbital shaking apparatus and drum apparatus are often needed for cell growth. To avoid contaminating both the cells and reactions, sterile disposable cell culture supplies are needed in great quantity. A CO_2 incubator is needed to grow mammalian cells at 37°C. A bacterial incubator is necessary. A sterile laminar flow cabinet for working with mammalian cells is helpful and may be necessary. For some recombinant studies, BCL2 or BCL3 level containment may be necessary. To visualize cells, a microscope (inverted light source) is needed.

Sterilization. An autoclave is necessary for sterilizing solutions, glassware, and other items for use in many of the methods. Filters and filtration units (0.22 μm) are used to sterilize solutions. Ultraviolet light, 70% ethanol, and flame are also utilized for sterilization of equipment.

Centrifuges. Most methods call for multiple centrifugations. A microcentrifuge, a tabletop centrifuge (refrigerated) for larger tubes, a floor model refrigerated centrifuge (with inserts for tubes of many sizes), and an ultracentrifuge (with fixed-angle, vertical, and swinging rotors) are required. A second microcentrifuge for use in the cold room is very handy. Because some procedures require long centrifugations, access to a few of each centrifuge type is helpful.

Refrigeration. A refrigerator, a −20°C freezer (can be part of a refrigerator), and access to a −70°C freezer are needed. Because many reagents are kept at −20°C, a separate freezer unit is useful. A cold room is useful for performing reactions and centrifugations at 4°C. Refrigerator/freezer space is a valuable commodity, and you will soon find yourself overcrowded, regardless of initial facility size. Liquid nitrogen, dry ice, and crushed ice are also necessary.

Safety. A fume hood is necessary because many solvents give off toxic fumes. Plexiglass shielding is needed to work with ^{32}P. Proper storage and waste containment systems are also needed. Thousands of plastic gloves are used (glove material will vary, depending on the specific need, the solvent used, or one's preference). Appropriate gloves for handling animals, dry ice, and glassware are also necessary. Your institution has rules governing the use and disposal of animals, chemicals and solvents, and radioactive materials. Depending on the nature of these rules, additional special equipment and supplies for storage and handling may be required.

Plasticware. Tubes of many different sizes are used in these methods. The most common are 1.5-ml microfuge tubes, polypropylene tubes of approximately 10 and 15 ml, snap-cap tubes in these sizes, 15- and 50-ml sterile screwcap tubes (commonly used in cell culture), and a variety of tubes for high-speed centrifugation. Disposable flasks of various sizes, culture dishes, microtiter plates, pipettes, and other cell culture–related plasticware are also used.

Glassware. Large flasks (e.g., 1–4 liters) for cell growth are needed, as well as a variety of glass reagent bottles. Thick-walled test tubes (e.g., 30-ml Corex tubes) are used for reactions and centrifugation. Common laboratory glassware (flasks, beakers, graduated cylinders, etc.) is also used. All glassware used for DNA or RNA samples should be acid washed and sterilized before use.

Radioactivity measurement. A liquid scintillation counter is necessary for work with ^{3}H, ^{14}C, ^{35}S and ^{32}P-radiolabeled material. A Geiger counter is very useful for monitoring the handling of ^{32}P-labeled radioactive substances and waste.

Light measurement. A UV/visible spectrophotometer is needed for many procedures. A refractometer is helpful, but not necessary. An ultraviolet (UV) transilluminating light source (302 nm) is used to visualize reaction products. A hand-held UV source, in addition to one used for photographing gels, is recommended.

Common laboratory equipment. A pH meter, hotplate-stirrer, vortex mixer, balances for weighing reagents, laboratory timers, pipettors, and desiccators are all required. A gradient maker and a fraction collector are needed. Sources of vacuum, nitrogen, air, CO_2, and O_2 are also needed.

Water. Of critical importance is a source of deionized, distilled H_2O of high purity. Note that all steps in this book calling for H_2O require the use of this pure water. Some procedures call for autoclaved or tap water; they are noted as such. Water of inferior purity is a potential source of problems.

EQUIPMENT SPECIFIC TO THESE METHODS

Photography. A Polaroid camera system to take pictures of gels is necessary, along with a light box transillumination system with a long UV source (approximately 302 nm).

Autoradiography. Cassettes to hold X ray film during autoradiographic exposure are used frequently. A system to develop X ray film is required. An automatic developer is very helpful, but not necessary. Image enhancing screens are useful for these procedures.

Electrophoresis systems. At least two power supplies and electrophoresis systems for vertical gels, horizontal gels, and minigels are needed. Power supplies delivering either high amperage (e.g., 400 mA) or high voltage (e.g., 2,000 V) are needed for some procedures. Glass and plastic plates, combs, spacers, clips, and so on are used with these systems.

Filters. Whatman 3MM paper filters in large sheets, 23-cm-wide rolls, and 20 × 20-cm squares are used. Nitrocellulose (NC) filters are used for blotting agar plates or gels. It is advisable to buy precut 20 × 20-cm sizes and 82-mm culture dish-size circles. (Some new filters such as nylon mesh are being used in place of nitrocellulose, and you may wish to try this substitution.) Other paper filters for filtration or scintillation counting are also used.

Dishes and pans. A variety of plastic dishes and pans are used to hold, store, and react gels, filters, and so on. Clear plastic food or storage dishes can be purchased in a department or discount store for these purposes.

Special equipment. A plastic bag heat-sealing system is invaluable for working with nitrocellulose filters. A variety of syringes and needles are needed for working with small volumes, radioactive samples, and other uses. Pipettes for very small volumes, repeating, and automatic pipettes are needed. A microwave oven is recommended for heating solutions, but is not necessary. A gel dryer is needed to dry sequencing gels. Also required is a vacuum drying system for desiccating samples. For transgenic mouse analysis, a number of special instruments are required (Section 21-1).

Homemade equipment. It is possible to use homemade equipment, such as electrophoresis chambers. Some items, such as a shock chamber for electroporation, may not be readily available and thus may have to be made. In general, with the growing number of suppliers of equipment and reagents for molecular biology, these items are readily available.

SECTION

4

Cloning Vectors and Bacterial Cells

SECTION **4-1.**

pBR322

This is a 4,362-bp double-stranded DNA plasmid cloning vehicle designed to allow simple and rapid preparation of cloned recombinant DNA fragments. It contains two antibiotic resistance genes (tetracycline and ampicillin), an origin of replication, and a variety of useful restriction sites for cloning or subcloning restriction fragments. pBR327 is identical to pBR322, except that it is missing all nucleotides between 1,427 and 2,516. These sequences were deleted to reduce the size of the cloning vector and to eliminate sequences that were shown to interfere with the expression of cloned DNA in eukaryotic cells. pBR327 still contains both antibiotic resistance genes. Nucleotide 1, by convention, is the first T in the *Eco*RI site G A A T T C. pBR322 and pBR322-derivative plasmids are very common plasmid vectors.

These plasmid cloning vectors confer antibiotic resistance when *E. coli*, deficient in the resistance genes, are successfully transformed. The two antibiotic resistance genes possess unique cloning sites for insertion of DNA fragments. Opening the plasmid with an RE and inserting a compatible DNA segment with DNA ligase inactivates this resistance gene. However, selection can be achieved by growing and plating the transformed cells in the presence of the other antibiotic. For example, if a cloned DNA is inserted into the *Pst*I site of pBR322, the plasmid will still confer tetracycline, but not ampicillin, resistance when successfully transformed to *E. coli*.

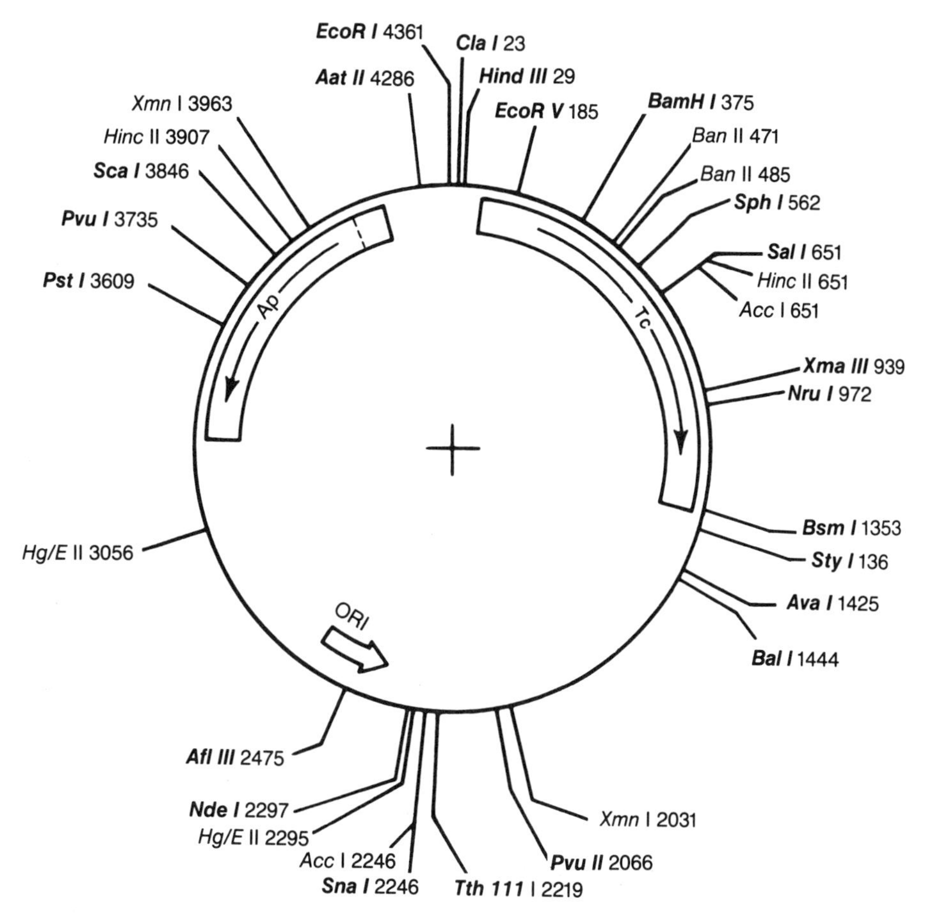

Figure 4.1
Restriction map of pBR322 plasmid. The REs in bold type indicate enzymes that have only one cleavage site. (Illustration courtesy of New England Biolabs.)

REFERENCES

Bolivar, F., Rodriguez, R., Greene, P., Betlach, M., Heynecker, H. and Boyer, H., Gene *2*:95, 1977.

Sutcliffe, J., Proc. Natl. Acad. Sci., USA *75*:3737, 1978.

SECTION **4-2.**

M13

M13 is a filamentous, male-specific, *E. coli* bacteriophage that contains a single-stranded circular DNA genome of 7.2 kb. The M13 mp series of cloning vehicle are modifications of the wild type M13 phage that allow easy cloning of a wide variety of DNA fragments into the bacteriophage vector. A polylinker sequence (or cluster of restriction sites) has been introduced into the amino terminus of the *lac Z* gene. The *lac Z* gene is the structural gene encoding the enzyme for metabolizing galactose sugars (β-galactosidase). Infection of some bacteria deficient in β-galactosidase (gal^-), such as JM103 or JM109, with a non-insert bearing M13 mp phage complements the host cell's gal^- phenotype, making these bacteria gal^+, and allows the bacteria to metabolize the galactose analog Xgal to a blue pigment. A cloned insert located in the polylinker region will interrupt the *lac Z* gene, rendering the bacteriophage gal^-. Thus some lac^- *E. coli*, such as JM103 or JM109, will be unable to metabolize Xgal into a blue pigment, even after infection with a recombinant M13 mp phage. Therefore, clear plaques will represent insert-bearing phage, and blue plaques will contain wild type nonrecombinants.

Single-stranded oligonucleotide primers are available from a variety of commercial sources and allow the nucleotide sequence of M13 mp inserts to be directly determined using the dideoxy chain termination method. In addition, primer extension allows the generation of radioactive single-stranded probes for S_1 nuclease analysis.

Starting with M13 mp8 and mp9, vectors have been constructed in pairs with a reversed order of RE sites in the cloning region. This is especially useful in sequencing from both ends of an insert with asymmetric termini. Two new vectors, mp18 and mp19, have recently been added to the mp series. They contain several new restriction sites in this polylinker region, allowing more flexibility in cloning. These bacteriophage are very useful in sequencing by the dideoxy chain termination method (Section 16).

REFERENCES

Messing, J., Gronenborn, B., Muller-Hill, B., and Hofschneider, P. H., Proc. Natl. Acad. Sci., USA, *74*:3642, 1977.

Messing, J. In: Methods in Enzymology *101* (part C): *Recombinant DNA*. (R. Wu, L. Grossman, and K. Maldave, eds.). Academic Press, New York, 1983, pp. 20–78.

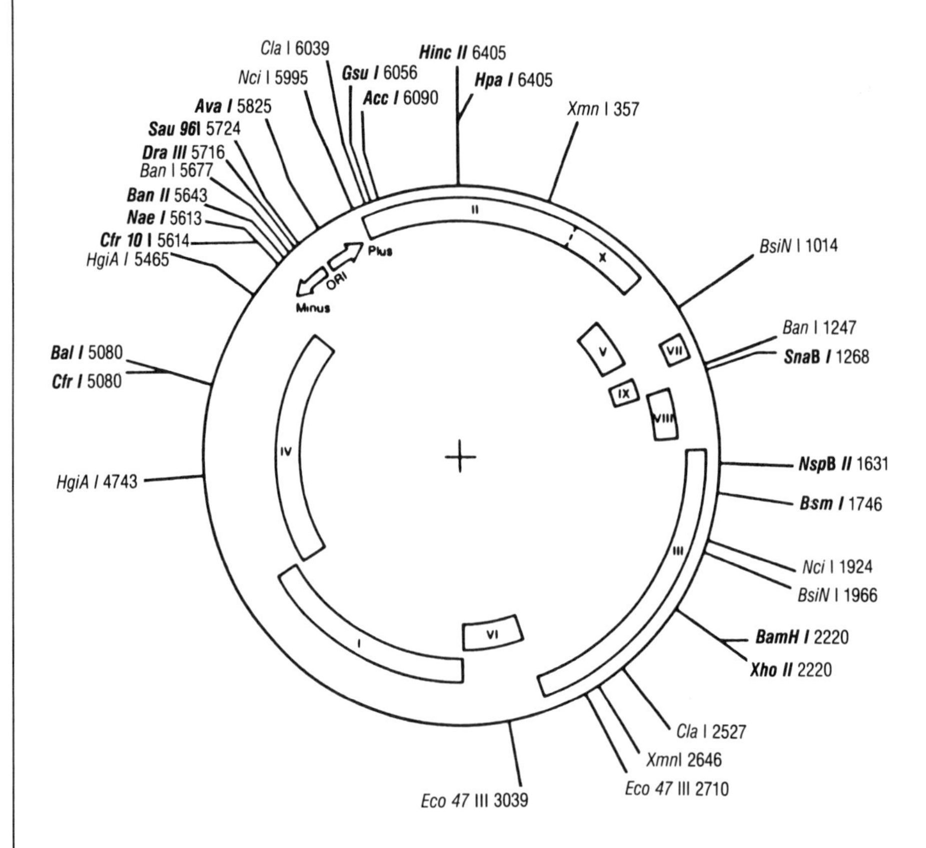

Figure 4.2
Restriction map of M13 wild type (above) and the multiple cloning sites of mp7, mp8, mp9, mp10, mp11, mp18 and mp19 constructs. The mp series of vectors are constructed in modified versions of wild type M13. (Illustrations courtesy of New England Biolabs and Pharmacia P-L.)

Multiple Cloning Sites

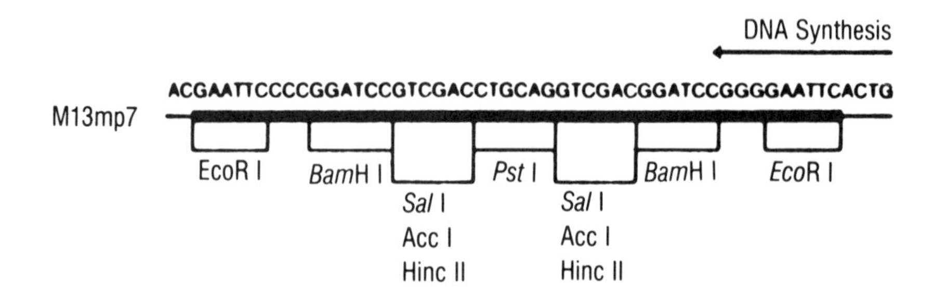

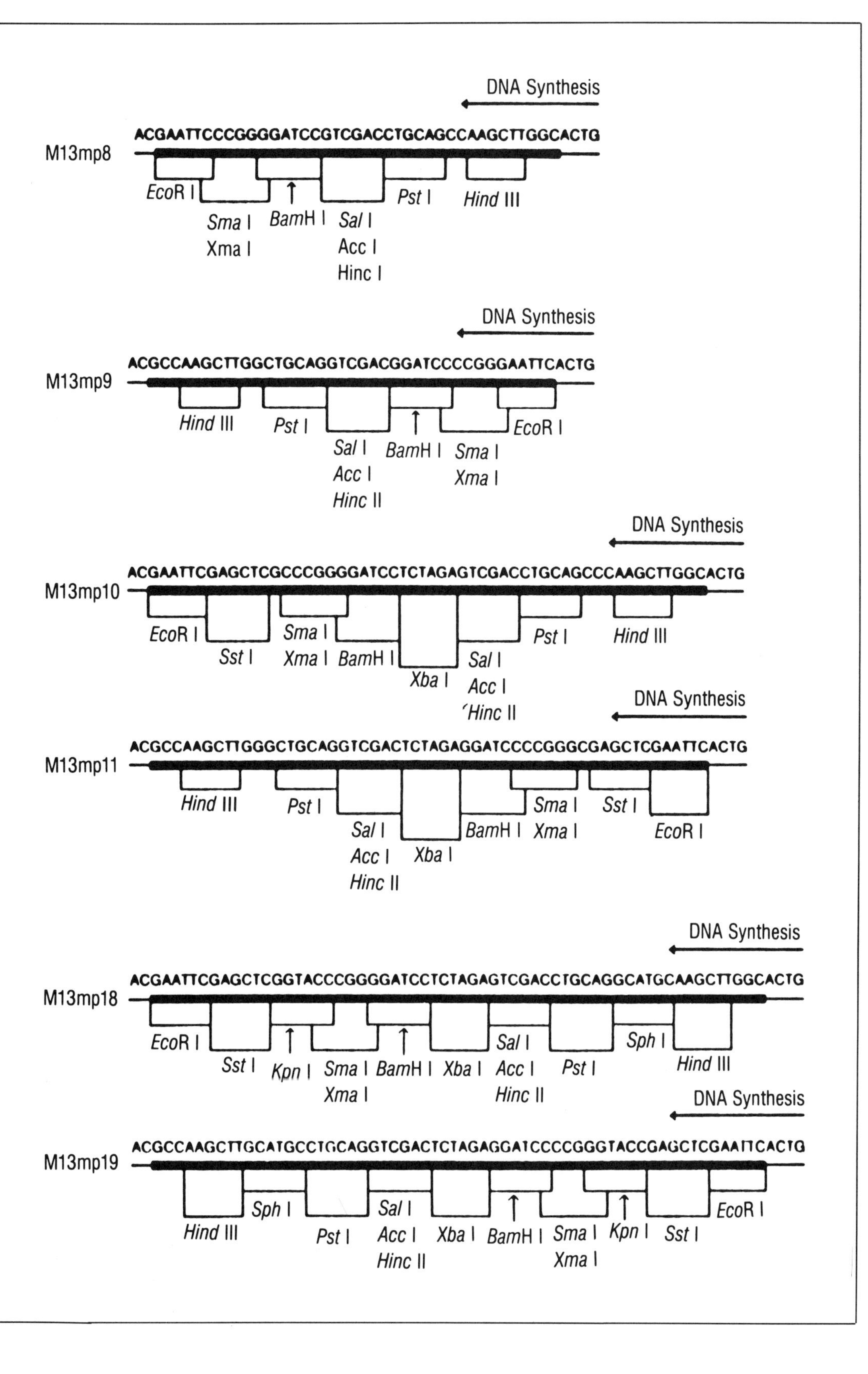
DNA Synthesis
ACGAATTCCCGGGGATCCGTCGACCTGCAGCCAAGCTTGGCACTG
M13mp8
EcoR I
Sma I
Xma I
BamH I
Sal I
Acc I
Hinc I
Pst I
Hind III
DNA Synthesis
ACGCCAAGCTTGGCTGCAGGTCGACGGATCCCCGGGAATTCACTG
M13mp9
Hind III
Pst I
Sal I
Acc I
Hinc II
BamH I
Sma I
Xma I
EcoR I
DNA Synthesis
ACGAATTCGAGCTCGCCCGGGGATCCTCTAGAGTCGACCTGCAGCCCAAGCTTGGCACTG
M13mp10
EcoR I
Sst I
Sma I
Xma I
BamH I
Xba I
Sal I
Acc I
Hinc II
Pst I
Hind III
DNA Synthesis
ACGCCAAGCTTGGGCTGCAGGTCGACTCTAGAGGATCCCCGGGCGAGCTCGAATTCACTG
M13mp11
Hind III
Pst I
Sal I
Acc I
Hinc II
Xba I
BamH I
Sma I
Xma I
Sst I
EcoR I
DNA Synthesis
ACGAATTCGAGCTCGGTACCCGGGGATCCTCTAGAGTCGACCTGCAGGCATGCAAGCTTGGCACTG
M13mp18
EcoR I
Sst I
Kpn I
Sma I
Xma I
BamH I
Xba I
Sal I
Acc I
Hinc II
Pst I
Sph I
Hind III
DNA Synthesis
ACGCCAAGCTTGCATGCCTGCAGGTCGACTCTAGAGGATCCCCGGGTACCGAGCTCGAATTCACTG
M13mp19
Hind III
Sph I
Pst I
Sal I
Acc I
Hinc II
Xba I
BamH I
Sma I
Xma I
Kpn I
Sst I
EcoR I

SECTION **4-3.**

pUC

This series of plasmids contains many features of M13 and pBR plasmids that are convenient for molecular biologists. These small plasmids (2.7 kb) have the pBR322 ampicillin resistance gene, the pBR322 origin of replication, and a portion of the *lac Z* gene of *E. coli.* Within the *lac* region is a polylinker sequence of unique RE recognition sites identical to those in the multiple cloning sites of M13. The pUC plasmids grow to a high copy number and can be amplified with chloramphenicol. When DNA fragments are cloned into this region of pUC, the *lac* gene is inactivated. These plasmids, when transformed into an appropriate *lac*$^-$ *E. coli,* such as JM103 or JM109, and grown in the presence of IPTG and Xgal, will give rise to white or clear colonies. Noninsert-containing pUC transformed into the same cells will yield blue colonies. As with the M13 series, pUC vectors are available in pairs with a reversed order of the RE sites in the cloning region relative to the *lac Z* promoter. pUC 8 and 9 and pUC 18 and 19 are two such pairs. The use of pUC vectors is described in Section 15-4.

REFERENCE

Vieira, J. and Messing, J., Gene *19*:259, 1982.

Figure 4.3
Diagram of pUC plasmids, indicating restriction maps of polylinker regions of pUC 8, 9, 12, 13, 18 and 19. (Illustration courtesy of Pharmacia P-L.)

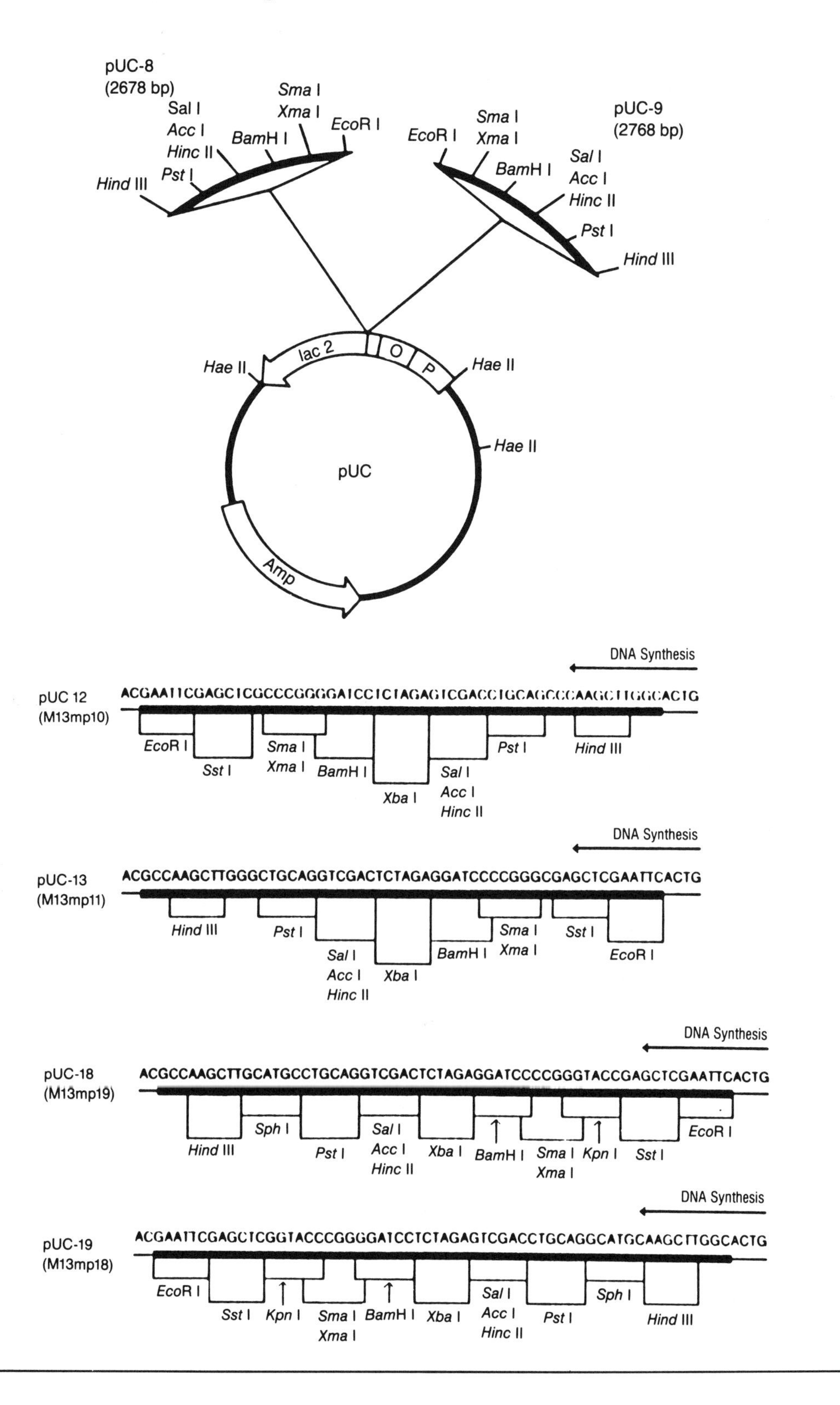
pUC Multiple Cloning Sites
pUC-8
(2678 bp)
Sal I
Acc I
Hinc II
Pst I
Hind III
BamH I
Sma I
Xma I
EcoR I
pUC-9
(2768 bp)
EcoR I
Sma I
Xma I
BamH I
Sal I
Acc I
Hinc II
Pst I
Hind III
lac 2
O
P
Hae II
Hae II
Hae II
pUC
Amp
DNA Synthesis
pUC 12
(M13mp10)
ACGAATTCGAGCTCGCCCGGGGATCCTCTAGAGTCGACCTGCAGCCCAAGCTTGGCACTG
EcoR I
Sst I
Sma I
Xma I
BamH I
Xba I
Sal I
Acc I
Hinc II
Pst I
Hind III
DNA Synthesis
pUC-13
(M13mp11)
ACGCCAAGCTTGGGCTGCAGGTCGACTCTAGAGGATCCCCGGGCGAGCTCGAATTCACTG
Hind III
Pst I
Sal I
Acc I
Hinc II
Xba I
BamH I
Sma I
Xma I
Sst I
EcoR I
DNA Synthesis
pUC-18
(M13mp19)
ACGCCAAGCTTGCATGCCTGCAGGTCGACTCTAGAGGATCCCCGGGTACCGAGCTCGAATTCACTG
Hind III
Sph I
Pst I
Sal I
Acc I
Hinc II
Xba I
BamH I
Sma I
Xma I
Kpn I
Sst I
EcoR I
DNA Synthesis
pUC-19
(M13mp18)
ACGAATTCGAGCTCGGTACCCGGGGATCCTCTAGAGTCGACCTGCAGGCATGCAAGCTTGGCACTG
EcoR I
Sst I
Kpn I
Sma I
Xma I
BamH I
Xba I
Sal I
Acc I
Hinc II
Pst I
Sph I
Hind III

SECTION **4-4.**

λgt10

λgt10 is a 43-kb linear double-stranded DNA λ bacteriophage-derived cloning vehicle that is useful for cloning small (<7-kb) *Eco*RI fragments such as cDNA inserts that have *Eco*RI linkers ligated to their ends. The insertion of DNA into the unique *Eco*RI site of λgt10 inactivates the cI (repressor) gene, generating a cI^- bacteriophage. Nonrecombinant λgt10 is cI^+ and forms a cloudy plaque on an appropriate *E. coli* host such as c600. Recombinant cI^- phage form clear plaques that are readily discernible from cloudy plaques. In addition, in an *E. coli* strain carrying the *hfl*A150 mutation (high frequency lysogeny mutation), only a cI^- bacteriophage will form plaques, because the cI^+ nonrecombinant phage will efficiently form stable lysogens and will not be visible as plaques. Thus, a strong selection exists against plaque formation by non-insert-bearing bacteriophages in c600 *hfl*A using the λgt10 system.

REFERENCES

Huynh, T., Young, R., and Davis, R., Constructing and screening cDNA libraries in λgt10 and λgt11. In: *DNA Cloning: A Practical Approach* (D. Glover, ed.) IRL Press, Oxford, 1984.

Murray, N., Brammer, W., and Murray, K., Mol. Gen. Genet. *15:*53, 1977.

Young, R., and Davis, R., Proc. Natl. Acad. Sci., USA *80:*1194, 1983.

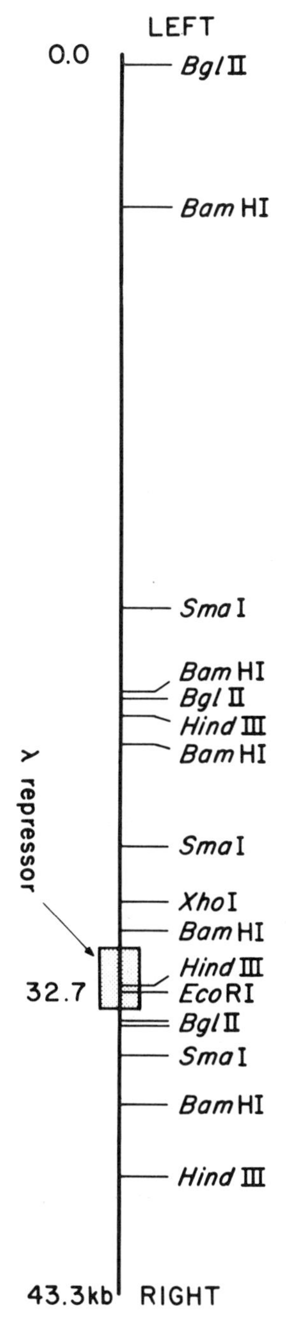

Figure 4.4
Diagram of λgt10, showing major RE sites.

SECTION **4-5.**

λgt11

λgt11 is a 43.7-kb linear double-stranded λ bacteriophage cloning vector designed exclusively for cloning small *Eco*RI fragments (less than 6 kb), in particular cDNA fragments with *Eco*RI linkers. The unique *Eco*RI site used for inserting foreign DNA is located near the COOH terminus of the *lac Z* gene. Foreign DNA sequences in this cloning vehicle can be expressed as β-galactosidase fusion proteins. Recombinant libraries generated from λgt11 can be screened with either antibody probes or nucleic acid probes, because the antibody may detect the fusion protein expressed by the recombinant λgt11 bacteriophage. In addition, insertion of foreign DNA into the *Eco*RI site renders the bacteriophage gal^-, whereas nonrecombinant λgt11 is gal^+. On an appropriate *E. coli* host that is gal^- (Y1090), recombinant phage will form clear plaques on Xgal plates, whereas nonrecombinants will form blue plaques. The presence or absence of inserts can therefore be easily assessed. In contrast to λgt10, however, there is no selection against nonrecombinant phage. Thus, for screening with nucleic acid probes, λgt10 is a better vector. For antibody probes, λgt11 is used.

This cloning vehicle has two other useful genetic properties: (1) it contains a temperature-sensitive repressor of lytic growth (*c*I857) that is inactive at 42°C and active at 32°C, allowing a selection against lysogenic growth, and (2) it has an amber mutation, *S*100, that renders the phage host lysis defective in all *E. coli* hosts that do not contain a strong amber supression mutation.

REFERENCE

Young, R., and Davis, R., Proc. Natl. Acad. Sci., USA *80:*1194, 1983.

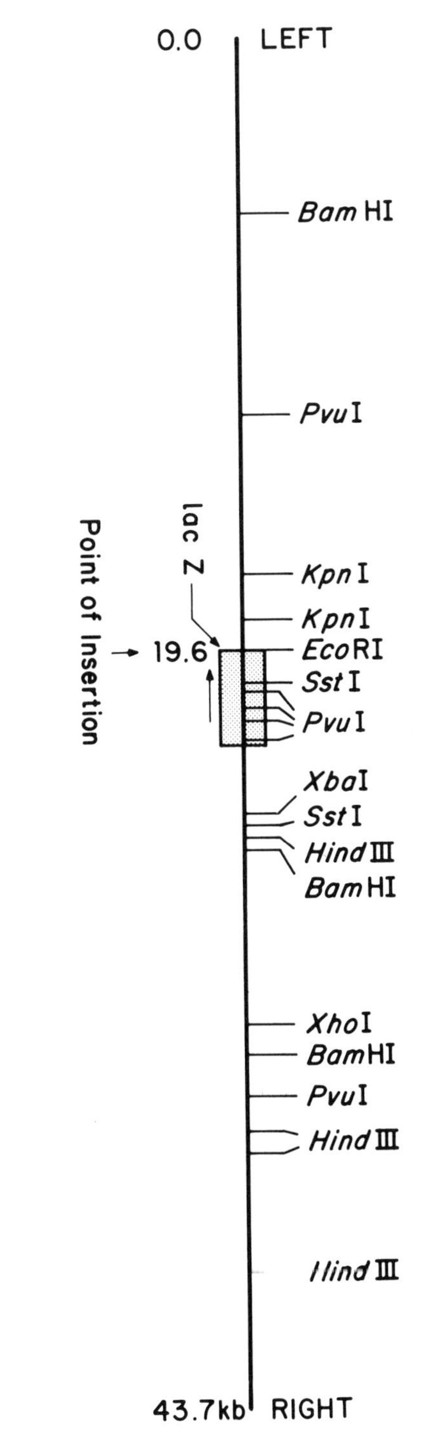

Figure 4.5
Diagram of λgtll, showing major RE sites.

SECTION **4-6.**

EMBL3 and EMBL4

EMBL3 is a λ replacement vector carrying two polylinker sequences that facilitate the cloning of *Mbo*I, *Bam*HI, *Eco*RI, *Bgl*II, *Xho*I, *Bcl*I, and *Sal*I cut fragments. It is most useful as a cloning vector for generating genomic libraries after partial *Mbo*I digestion. This cloning vector can accomodate up to approximately 23-kb inserts, making it suitable for genomic library construction. The design of the polylinker sequences simplifies preparation of the vector for cloning partial *Mbo*I fragments 15–20 kb in size. In addition, the polylinker sequences allow easy removal of cloned DNA at the cloning site. This double-stranded, 44-kb DNA bacteriophage grows well on appropriate *E. coli* host strains such as LE392. It is a useful vector for genomic library construction. A second vector, EMBL4, is available with RE sites in reverse order in the polylinker region for cloning *Eco*RI fragments.

REFERENCE

Frischauf, A.-M., Lehrach, H., Poustka, A., and Murray, N., J. Mol. Biol., *170:*827, 1983.

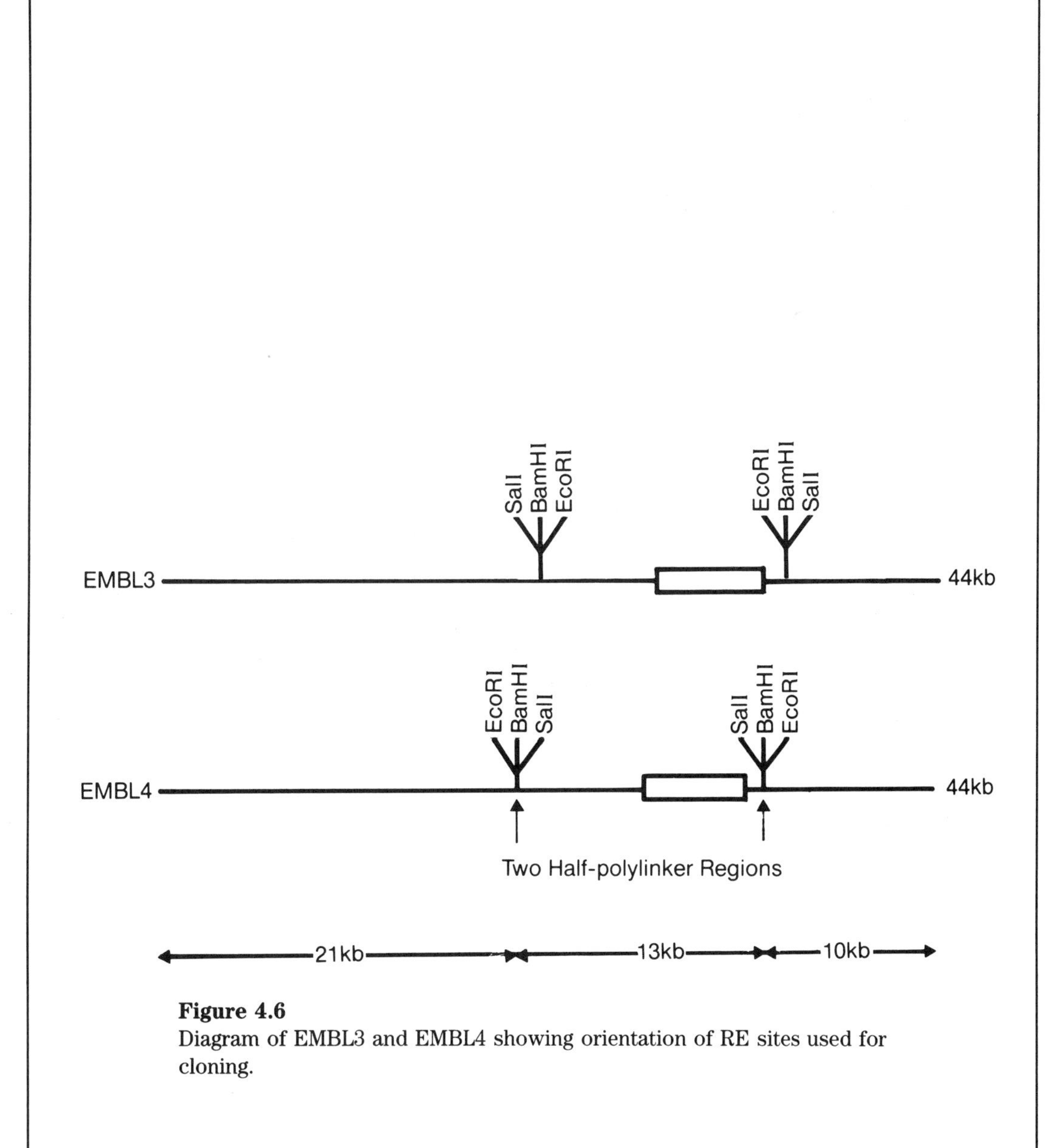

Figure 4.6
Diagram of EMBL3 and EMBL4 showing orientation of RE sites used for cloning.

SECTION

4-7.
Charon 28

Charon 28 is one of the original λ replacement vectors designed to clone large *Bam*HI and *Mbo*I fragments. This cloning vehicle can accomodate up to 20-kb inserts. Many eukaryotic genomic libraries in existence today have been made with this vector. It lacks the polylinker at the insertion sites that is present in EMBL3, making excision of a cloned insert or preparation of the cloning vector somewhat more difficult. In addition, the plaque size of various recombinants varies greatly in libraries constructed of Charon 28, a disadvantage when the library is screened.

REFERENCE

Blattner, F., et al., Science, *196:*161, 1977.

Rimm, D. L., Horness, D., Kucera, J., and Blattner, F. R., Gene *12:*301, 1980.

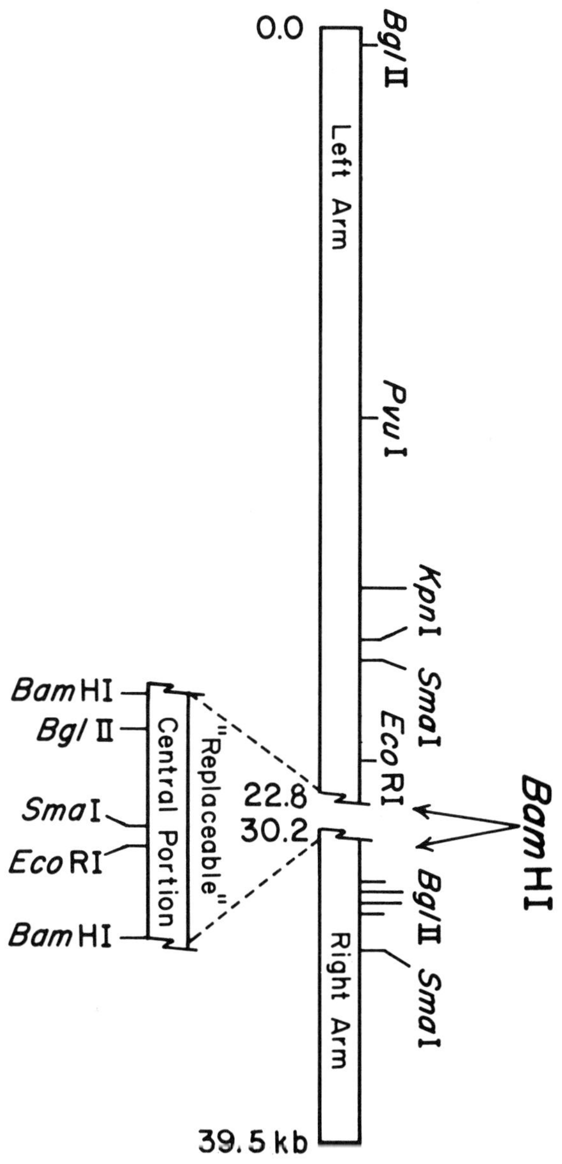

Figure 4.7
Map of representative RE sites on Charon 28 #2221 vector. Charon 28 is a replacement vector and the gap denotes central portion removed for acceptance of DNA from 5 to 20 kb in length.

SECTION

4-8. Bacterial Strains

Most host bacterial cells are *E. coli* derived. Because of their unique genetic properties, different cloning vectors require specific host cells for transfection. The table below lists the host cells used with vectors described in this manual.

Vector	Host Cell*	Growth Medium	Section
pBR322	LE392, HB101, JM107, JM109	LB/antibiotic	15–2
pUC series	JM103	LB/ampicillin	15–4
bacteriophage λ	LE392	LB with $MgSO_4$	12–1
M13	JM103, JM107, JM109	LB	16–5
EMBL3	LE392	LB with $MgSO_4$	13–5
Charon 28	LE392, c600	LB with $MgSO_4$	13–5
λgt10	c600, c600 *hfl*A	LB with $MgSO_4$	14–1
λgt11	Y1090, BNN97	LB/ampicillin with $MgSO_4$	14–1
pSP64, pSP65	LE392, HB101	LB/ampicillin	11–7
Bluescribe	JM109	LB/ampicillin	11–7

* Stocks of JM103, JM107, and JM109 are maintained on minimal agar plates.

There are alternative bacterial hosts for each vector system that may be suggested by individual suppliers. Bacterial strains can be purchased from the American Type Culture Collection (ATCC) and other cell repository agencies (see Appendix III).

SECTION

5

Preparation of DNA from Eukaryotic Cells

SECTION

5-1. Rapid DNA Preparation

DESCRIPTION

This method allows for preparation of many small samples with a minimum of steps and time, generating DNA whose quality is adequate for Southern blot analysis of restriction fragments. The purified DNA is not good enough in quality for a DNA library because the Polytron shears the genomic DNA.

TIME REQUIRED

4–5 hr

SPECIAL EQUIPMENT

Polytron or equivalent
Shaker

REAGENTS

Homogenization buffer: 4 ml

- **0.1 M NaCl**
- 0.2 M Sucrose
- **0.01 M EDTA**
- 0.3 M Tris, pH 8.0

10% SDS
SS-phenol
Chloroform
8 M potassium acetate

Ethanol
TE buffer

METHODS

1. Place tissue DNA source (e.g., mouse tail piece, under 100 mg) in a 10-ml snap-cap polypropylene tube.[1]
2. Add 2 ml of homogenization buffer.
3. Grind with tissue homogenizer (e.g., Polytron) until no pieces are visible.
4. Add 125 μl of 10% SDS and mix by vortexing.
5. Incubate in water bath at 65°C for at least 30 min.
6. Add 350 μl of 8 M potassium acetate and mix by vortexing. Incubate sample on ice for 60 min.
7. Spin in centrifuge at 5,000 × g for 10 min at 4°C.
8. Transfer supernatant to a new 10-ml tube. Discard pellet.
9. Add 2 ml of chloroform and 2 ml of SS-phenol to extract. Mix by vortexing briefly or inverting tube. Let stand or centrifuge briefly until phases separate.
10. Remove the upper aqueous layer to a new tube. Discard the lower layer.
11. Add 2 ml of chloroform. Mix as above. Let stand until phases separate. Remove lower chloroform phase and discard.
12. To the aqueous layer, add 5 ml of ethanol to precipitate DNA. Mix gently by inverting tube. Spin for 10 min at 1,500 × *g*. Precipitated DNA will form a pellet.
13. Remove supernatant and discard. Add 5 ml of 80% ethanol. Mix gently.
14. Spin for 5 min at 1,500 × *g*.
15. Remove ethanol. Drain off liquid and let tube stand at room temperature for at least 30 min to dry pellet partially. Do not dry completely.
16. Resuspend pellet in 300 μl of TE buffer. Shake gently at room temperature for 90 min. Store at 4°C. DNA sample is stable and will keep for over 6 months.
17. Sample is now ready for RE cleavage (Section 5-4) and loading onto agarose gel for electrophoresis (Section 5-5).[2]

---

1 A good tissue source is the 2-cm tip of mouse tail. This is particularly useful when genomic DNA analysis is needed on a live mouse before future work. A DNA source can also be *Drosophila*, cultured cells, or small pieces of fresh or frozen tissue. Tissues with high levels of endogenous DNases may yield degraded DNA samples.

2 Because of RNA contamination in a sample, accurate O.D. measurements of DNA content cannot be made from DNA prepared by this method. A 30 μl aliquot typically contains 10 μg of DNA.

SECTION **5-2.**

Preparation of DNA from Eukaryotic Cells: General Method

DESCRIPTION

This method provides a high-quality preparation of high molecular weight genomic DNA that is suitable for Southern blotting and library construction. Proteinase K is used to liberate nucleic acids from cells, followed by removal of RNA by RNase A. This method has been used primarily with cultured cells.

TIME REQUIRED

Day 1—1 hr
Day 2—5 hr
Day 3—6 hr

REAGENTS

Phosphate-buffered saline (PBS)
Proteinase K solution, 1 mg/ml (BM) in:
- **10 mM Tris, pH 7.4**
- **10 mM EDTA**
- **150 mM NaCl**
- **0.4% SDS** (add last)

SS-phenol
Chloroform
TE buffer
RNase A (DNase free), 10 mg/ml[1]

---

[1] DNase-free RNase A can be made by suspending RNase A at 10 mg/ml of TE buffer and heating for 10 min at 70°C to inactivate DNase, followed by slow cooling.

3 M sodium acetate, pH 7.4
Ethanol

METHODS

In Advance

Grow cells for DNA extraction. Tissue culture cells of various types work well.

DNA Extraction

1. Centrifuge up to 10^9 cells in a 15-ml capped tube at 1,000 × g for 5 min.[2] Decant medium.
2. Resuspend pellet in 10 ml of PBS. Repeat step 1.
3. Mix 1 volume of packed cells with 10 volumes of proteinase K solution.
4. Incubate for 15 min at 65°C.
5. Shake gently overnight at 37°C in a 50-ml polypropylene screw-cap tube.

Next Day–Clean DNA

6. Extract solution with an equal volume of 1:1 SS-phenol:chloroform. Mix well by inversion of the tube. Transfer to a 30-ml Corex tube. Separate phases by centrifugation for 10 min at 12,000 × g.
7. Remove the upper aqueous phase to a new tube with the large-opening end of a pipette. Be careful to leave the interface.
8. Repeat organic extraction steps 6 and 7 until interface material between phases is no longer visible.
9. Transfer aqueous phase to a new 30-ml tube.
10. Measure volume of aqueous phase. Add 1/10th of 1 volume of 3 M sodium acetate. Add 2.5 volumes of ethanol. Mix by inverting the tube.
11. Genomic DNA should precipitate immediately.
12. Centrifuge for 10 min at 12,000 × g to collect precipitated DNA. Wash pellet in 5 ml of 80% ethanol. Collect the pellet by centrifugation. Decant the ethanol and air-dry the pellet briefly.
13. Redissolve DNA pellet in 4.5 ml of TE buffer. Allow pellet to dissolve overnight at 37°C with shaking, if DNA is difficult to resuspend.

---

2 All tubes must be extremely clean. Wash 30-ml thick-walled glass tubes (e.g., Corex) with 50% nitric acid before using to remove any DNA from tubes. Rinse thoroughly and autoclave before use.

Next Day

14. Add 25 μl of 10 mg/ml RNase A (DNase free). Incubate for 30 min at 37°C.
15. Add 500 μl of 3 M sodium acetate. Add 5 ml of 1:1 SS-phenol:chloroform. Mix by inverting the tube many times.
16. Separate phases by centrifugation at 12,000 $\times$ g for 10 min.
17. Remove the upper aqueous phase to a new tube. Add 5 ml of chloroform. Repeat step 16.
18. Remove the upper aqueous phase to a new 30-ml Corex tube.[3] Add 12.5 ml of ethanol (2.5 volumes). Genomic DNA will precipitate immediately after solution is mixed by inverting the tube.
19. Spin as in step 16. Decant supernatant.
20. Add 5 ml 80% of ethanol to rinse pellet. Spin as in step 16.
21. Decant ethanol. Allow pellet to air-dry. Do not dry completely.
22. Resuspend DNA pellet in TE buffer. Large amounts of DNA may require overnight incubation at 37°C to dissolve completely.[4]
23. Read O.D. at 260 nm to estimate concentration (Section 20-3).
24. DNA can now be used for RE digestion and other characterization.[5]

---

[3] See note 2.

[4] Genomic DNA is difficult to resuspend in aqueous solution following ethanol precipitation. Resuspension may require overnight incubation at 37°C with shaking. It is best not to concentrate DNA at this step (i.e., keep DNA at less than 1 mg/ml) in order to achieve good dissolution of the pellet.

[5] If DNA does not cut well with REs, the preparation may be salvagable by repeating proteinase K digestion, organic extractions, and ethanol precipitation.

SECTION **5-3.**

DNA Preparation from Cultured Cells and Tissue

DESCRIPTION

This method is used to prepare genomic DNA from tissue or cultured cells with excellent purity. This method yields very high molecular weight DNA and involves an initial preparation of nuclei. This is the most involved and time consuming of the three methods.

TIME REQUIRED

Day 1—6–7 hr
Day 2—1–2 hr

REAGENTS

5% citric acid
0.88 M sucrose in 5% citric acid
1× RSB buffer
  10 mM Tris, pH 7.4
  10 mM NaCl
  25 mM EDTA

10% SDS
Proteinase K, powdered form (BM)
Diethyl ether
SS-phenol
Chloroform
Isoamyl alcohol
Ethanol
5 M NaCl

RNase A (BM) 10 mg/ml stock, made DNase-free by heating for 10 min at 70°C.

TE buffer

3 M sodium acetate, pH 7.4

Tris/salt buffer

10 mM Tris, pH 7.4

25 mM EDTA

0.5 M NaCl

0.1× SSC buffer

METHODS

In Advance

Prepare fresh tissue of interest for DNA isolation.[1] Blot off blood and dissect out connective tissue where necessary. Up to 5 g of tissue can be used in this preparation. Alternatively tissue can be frozen and stored at −70°C before initiating step 1.

To Prepare Tissue[2]

1. Pulverize, with a pestle, approximately 5 g of tissue in a mortar partially filled with liquid nitrogen (wear thermal gloves and eye protection). When tissue is finely ground to a powder, pour into 50-ml capped tube. Let liquid nitrogen evaporate. Add 10 ml of 1× RSB buffer. Go to step 2.

Or

To Prepare Nuclei

Dounce homogenize 5 g of freshly dissected tissue in 30 ml of 5% citric acid with a loose-fitting pestle until pestle moves freely. Then rehomogenize with a tight-fitting pestle until pestle moves freely up and down (8–10 times). Pour through 10 layers of sterile gauze (prewet with 5% citric acid). Centrifuge for 5 min at 2,500 × *g*. Suspend pellet in 10 ml of 5% citric acid by gentle manipulation with a glass pipette. Overlay homogenate on a 15-ml cushion of 0.88 M sucrose (30%) in 5% citric acid in an acid-washed 30-ml Corex tube. Spin at 5,000 × *g* for 5 min in a swinging bucket rotor (e.g., HB4 rotor at 3,500 rpm). Remove upper layers by aspiration. Resuspend nuclear pellet nuclei in 10 ml of 1× RSB buffer. Centrifuge at 2,500 × *g* for 2 min. Repeat wash of pellet in 1× RSB buffer until the suspension is pH 7.4 (test by placing a small drop on pH indicator paper).

---

[1] Most tissues of interest will work with this method. Comparison of various tissues may show DNA rearrangement, especially in some tissues (e.g., tumors or B cells). In general, 5 g of tissue should yield 1–5 mg of DNA.

[2] Frozen pulverization is better for most tissues and is an easier method, but dounce homogenization may prove better for working with cultured cells.

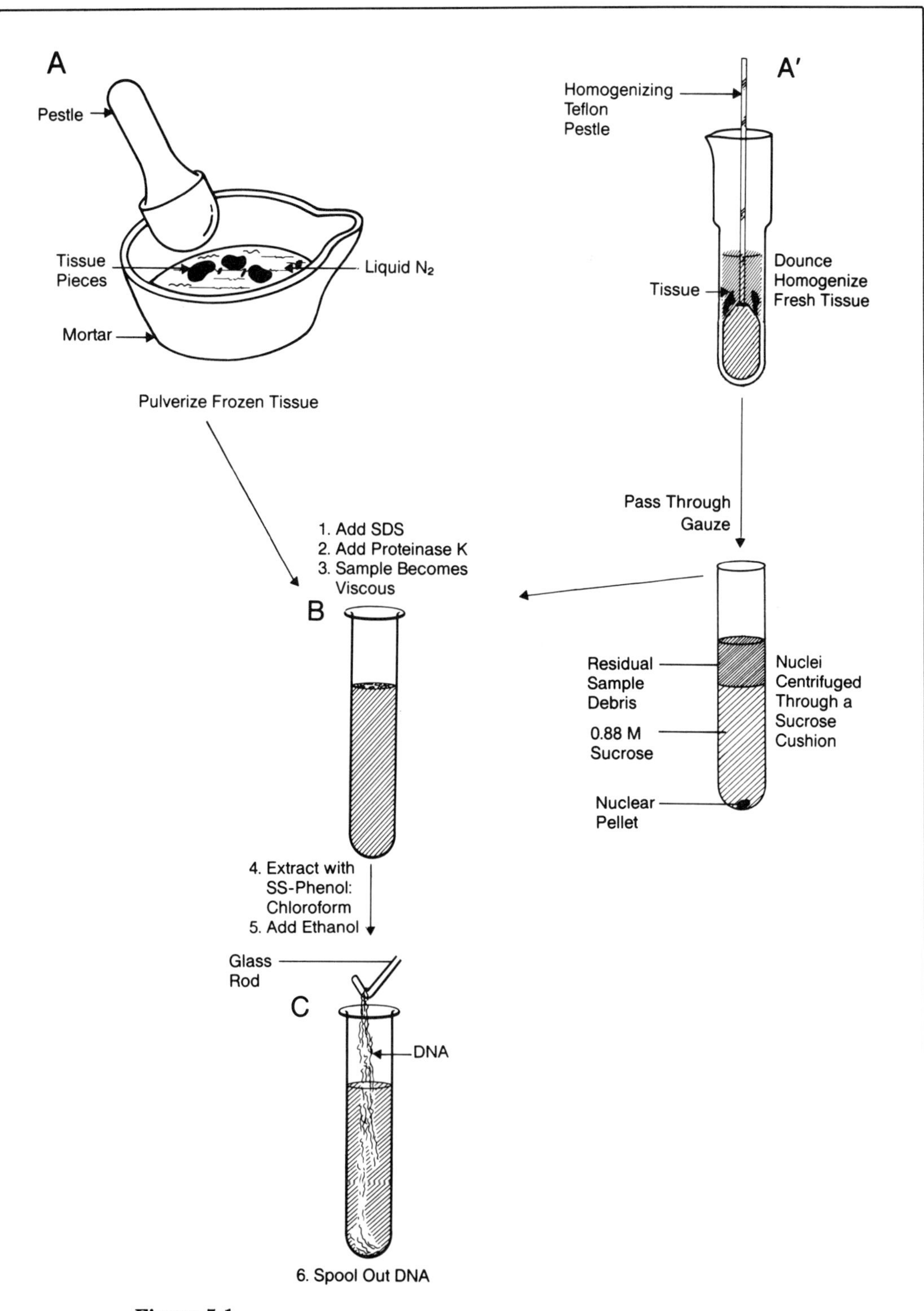

Figure 5.1
Preparation of DNA. Tissue can be frozen and pulverized (A) or nuclei separated by sucrose density centrifugation (A′). The DNA is separated from protein and cellular debris, extracted, and precipitated. Large molecular weight species can be removed with a glass rod.

2. Add SDS to pulverized tissue mixture or nuclei suspension to equal 1% SDS (1 ml of 10% SDS added is typical). Sample will become viscous as DNA is released from the nuclei.
3. Add 11 mg proteinase K powder (1 mg/ml concentration). Incubate for 1–2 hr at 37°C. Add another 11 mg of proteinase K and continue incubating for at least 1–2 hr more, or longer, until tissue is completely dissolved.
4. Add 1 ml of 5 M NaCl.
5. Extract with an equal volume of SS-phenol:chloroform:isoamyl alcohol (10:10:1). Mix gently but thoroughly by inverting tube. Briefly spin to separate phases. Save upper aqueous phase by transfering to a new tube using the large open end of a pipette.
6. Extract aqueous phase with 2 volumes of diethyl ether.[3]
7. Aspirate off ether (on top) and save lower aqueous phase.
8. Add 2.5 volumes of absolute ethanol to precipitate DNA.[4] Large pieces of DNA can be removed by collecting fibers with a glass rod and transferring them to a new tube. Excess ethanol is removed with a sterile cotton swab. If removal of RNA is desired, go to step 10.
9. Dissolve pellet in 5 ml of TE buffer overnight at 4°C. DNA is very stable in TE buffer at 4°C.[5] Go to step 16 unless further removal of RNA from the DNA preparation is desired.[6]

For Removal of Residual RNA

10. DNA can be purified further by dissolving in 10 ml of 0.1× SSC overnight at 4°C.
11. Add 20 μl of DNase-free RNase A. Incubate for 30 min at 37°C.
12. Add 1 ml of 5 M NaCl.
13. Add 10 ml of SS-phenol and 10 ml of chloroform. Mix gently by inverting tube. Centrifuge at 12,000 × g for 10 min to separate phases.
14. After phase separation, remove upper aqueous layer with the large open end of a pipette and place in a fresh tube.
15. Add 25 ml of ethanol (prechilled to −20°C). Remove DNA as in step 8. Dissolve pellet in TE buffer overnight at 4°C.[6]
16. Store samples at 4°C in TE buffer.

---

[3] Use in a fume hood.

[4] DNA should be a fibrous white material.

[5] Add 1 drop of chloroform to the sample for longer-term storage.

[6] The DNA solution may be too viscous to dissolve completely. If so, dilute with additional TE buffer.

SECTION **5-4.**

Restriction Endonucleases (REs) and Their Use

DESCRIPTION AND GENERAL NOTES

Restriction endonucleases (REs) are bacterial enzymes that cleave double-stranded DNA. Type I REs are important in bacterial function but do not cleave DNA at specific sequences. Type II REs, described for use in this manual, require highly specific sites for DNA cleavage and are thus extremely useful tools in molecular biology. These enzymes allow the cloning and purification of defined DNA fragments. The 500 or so known REs are typically isolated from a variety of bacterial strains.

REs are present in bacteria presumably to destroy DNA from foreign sources (e.g., infecting bacteriophage) by cleaving the foreign DNA at specific recognition sites. The host bacteria's DNA is protected from cleavage because the specific recognition sites are modified, usually by methylation at one of the bases in the site, making the site no longer a substrate for RE cleavage. Practically, REs with different recognition site specificities have been purified from various bacterial strains and are used by molecular biologists under defined conditions to cleave purified DNA from eukaryotic sources into defined fragments in an in vitro reaction. Host bacteria used to propagate cloned DNA in the laboratory are usually mutant in the host restriction genes (*hsdR*, *hsdM*, or *hsdS*); thus their intracellular enzyme activities will not destroy the foreign recombinant sequences.

The cleavage site specificities for many REs have been defined. For example, the RE *Eco*RI requires that six base pairs occur in the following specific order:

5′ . . . GAATTC . . . 3′
3′ . . . CTTAAG . . . 5′

*Eco*RI recognizes this sequence and cleaves it in a unique fashion, resulting in two termini with protruding 5′ ends:

. . . G   AATTC . . .
. . . CTTAA   and   G . . .

These ends are complementary ("sticky") and can be enzymatically reattached to any other *Eco*RI-generated termini using T4 DNA ligase. Other REs recognize different sequences and make unique cleavages resulting in other defined termini, some of which also have 5′ protruding ends, 3′ protruding ends, or no protruding (blunt) ends (see Table 5.1). Each RE has a specific sequence and number of nucleotides required to create the recognition site. Some REs do not require a specific nucleotide in every position of the recognition site. For example, in some cases a purine nucleotide (A or G) in a defined position is sufficient. The frequency with which an RE will cut DNA is thus related to the number of bases in the recognition sequences (if more bases are required, fewer cuts are apt to be made) and the specific bases in any position.

REs are useful because their specificity and the resulting ligatable termini allow the dissection, analysis, and restructuring of DNA in a controlled, predictable, site-specific manner. Because they are enzymes, each has specific requirements for conditions leading to optimal cleavage activity. Fortunately, many are active under similar conditions, with the primary variable being ionic strength (i.e., NaCl in the buffer). Practical conditions have been empirically defined for using each RE as a research tool.

The amount of RE activity is not usually defined using classical enzyme kinetics. Rather, RE activity is defined in practical terms for use in the laboratory. One unit (U) of activity for an RE is usually defined as the amount of enzyme that will cut 1 μg at all specific sites in a DNA sample (usually bacteriophage λ) in 1 hr at 37°C. The actual activity achieved may be different if additional RE digestion sites are present. For example, if 1 U of an RE is sufficient to cut 1 μg of bacteriophage λ DNA completely at one site under standard assay conditions, it may not be able to cut completely 1 μg of a DNA sample that has four sites for an RE, or a sample that has not been sufficiently purified.

Other factors concerning the reaction conditions also affect RE activity. These include purity of the DNA, buffer, temperature conditions, and the RE itself. DNA methylation, bound protein, or excessive viscosity of high molecular weight DNA in viscous solutions may decrease the efficiency of an RE digest. The DNA to be cut should usually be free of impurities; SS-phenol/chloroform extraction and ethanol precipitation (Section 20-1) will remove many interfering impurities. In contrast, most REs are usually not adversely affected by RNA or single-stranded DNA. Thus, the addition of tRNA as a coprecipitant can be useful in recovering small amounts of DNA without affecting subsequent RE digests.

The typical assay condition for an RE reaction contains a buffer (10 mM Tris, pH 7.4 is common), 10 mM Mg^{2+}, and is free of substantial concentrations of chelating agents (TE buffer contains a sufficiently low amount of EDTA). The RE buffers also contain proper salt concentrations to give the optimal ionic strength for each RE. In some cases, BSA, DTT, and spermidine are added to provide optimal enzyme activity. Four generalized RE buffers can satisfy the requirements for most REs, and the advance preparation of 10× stocks of these buffers is described below. Most RE digests are performed at 37°C, although there are a few exceptions, such as *Taq*I.

Most purchased REs are shipped with an appropriate storage buffer, but this contains 25–50% glycerol to prevent freezing during storage at −20°C. Concentrations of glycerol of over 5% (vol/vol) in RE digestion assays can inhibit the activity of many REs. In addition, high glycerol concentrations during RE digestion can relax the sequence specificity of some REs, resulting in spurious cleavages. For example, this phenomenon is seen with the commonly used RE *Eco*RI, where the spurious activity observed under some conditions is called *Eco*RI* (*star*) *activity*. For these reasons, it is important to make the volume of the RE digestion reaction large enough to encompass at least a 1:10 dilution of the RE itself (and thus reduce the glycerol concentration). Other factors, such as pH, ethanol, or improper salt concentration can also cause star activity or decreased activity; thus, careful adherence to recommended conditions is important.

A few other considerations are useful when using REs. Use a new pipette tip each time an RE is transferred. Trace amounts of contaminant added to the RE stock can prevent its further use and confound results obtained in subsequent experiments. When optimizing reactions, it is better to add additional RE than to incubate for much longer than 1 hr. Furthermore, as mentioned above, enzyme activity is defined only for cutting a DNA standard. Because your sample is generally quite different from the standard, a good guide is to use about 2–3 U of RE per microgram of DNA to be cut, especially when attempting to digest a difficult sample. It may sometimes be necessary to titrate the amount of RE needed to digest a DNA sample appropriately.

Finally, if two REs are to be used, attention must be given to the optimal salt conditions of each. If both require the same buffer, digestion can be done either simultaneously or sequentially. If different salt conditions are required, the RE requiring less salt must be used first. After incubation adjust the salt condition, and digest with the RE requiring high salt. RE activity can be removed by SS-phenol/chloroform extraction in all cases. Some REs are thermolabile and can be inactivated by heating to 70°C for 5 min. This property, however, varies from one RE to the next and should be verified for each RE used.

TIME REQUIRED

2 hr

REAGENTS

The buffer used depends on the optimal assay conditions for each RE. Enzyme suppliers recommend the best buffer to use and the activity of endonuclease. Consult the manufacturer's literature. A general list of commonly used REs and buffers is in the table.

Typical Stock Buffers

10× buffer O—no salt

100 mM Tris, pH 7.4	1 mg/ml BSA
100 mM $MgCl_2$	**10 mM DTT**

Table 5.1. Restriction Endonucleases

RE Name	Buffer Used	Specific Digestion Site
*Ava*I	L	5′ C YCGR G 3′ 3′ G RGCY C 5′
*Ava*II	L	G G$^{A}_{T}$C C C C$^{T}_{A}$G G
*Bal*I	O	TGG CCA ACC GGT
*Bam*HI	H	G GATC C C CTAG G
*Bgl*II	H	A GATC T T CTAG A
*Cla*I	L	AT CG AT TA GC TA
*Eco*RI	H	G AATT C C TTAA G
*Hae*III	L	GG CC CC GG
*Hinc*II	L	GTY RAC CAR YTG
*Hind*III	L	A AGCT T T TCGA A
*Hinf*I	H	G ANTC CTNA G
*Hpa*I	L or K	GTT AAC CAA TTG
*Hpa*II	O	C CG G G GC C
*Kpn*I	O	G GTAC C C CATG G
*Mbo*I	H or L	GATC CTAG
*Msp*I	L	C CG G G GC C

RE Name	Buffer Used	Specific Digestion Site
*Pst*I	H or L	C TGCA G G ACGT C
*Pvu*I	H	CG AT CG GC TA GC
*Pvu*II	L	CAG CTG GTC GAC
*Sac*I (*Sst*I)	O	G AGCT C C TCGA G
*Sac*II	L	CC GC GG GG CG CC
*Sal*I	H	G TCGA C C AGCT G
*Sau*3A	L	GATC CTAG
*Sma*I	K	CCC GGG GGG CCC
*Stu*I	H	AGG CCT TCC GGA
*Taq*I	L	T CGA AGC T
*Xba*I	H	T CTAG A A GATC T
*Xho*I	H	C TCGA G G AGCT C
*Xma*I	O	C CCGG G G GGCC C

Key: A—adenine
C—cytosine
G—guanine
N—any base
T—thymine
R—any purine
Y—any pyrimidine

Buffers: O—no salt
L—low salt
H—high salt
K—potassium salt

10× buffer L—low salt

As in buffer O with 0.5 M NaCl

10× buffer H—high salt

As in buffer O with 1.0 M NaCl

10× buffer K—potassium buffer

As in buffer O, except that Tris has a pH of 8.0 and 200 mM KCl is added.

METHODS

In Advance

Conditions vary depending on which RE is used and the purpose. Table 5-1 gives the cleavage site and suggested buffer condition for some typical REs. Note that all cleave double-stranded DNA and are highly specific for the indicated sites. Restriction maps of commercially available plasmids and bacteriophage vectors from suppliers will help in choosing the appropriate RE.

Analytical Digest

1. To run a small-volume RE reaction, incubate the following for 1 hr at 37°C in a 1.5-ml microfuge tube

   1 μl of appropriate 10× buffer
   1 μl of DNA sample (0.1–5 μg)
   1 μl of RE (containing 0.2 to 10 U)
   H_2O to make 10 μl

2. After incubation, cleaved DNA can be run on a minigel with standards, and an aliquot of the original uncut DNA (Section 9-1). This determines whether the RE cleaved the DNA to completion.[1] If the small-scale reaction worked well, the RE reaction can be rerun on a larger scale (e.g., in a 250- to 400-μl volume) to prepare DNA for further purification by electroelution from a preparative agarose or acrylamide gel (Section 9-3).

3. One may desire to calculate the RE concentration for a larger-scale preparation to determine the RE concentration:

   One unit of RE will cut 1 μg of DNA at 37°C over 1 hr. Typically, one should use 2–3 times as many units as are needed to ensure complete digestion.[2]

---

1 An aliquot of RE-digested DNA can be incubated with DNA ligase to reattach fragments before running a minigel. This procedure should generate high molecular weight concatameric DNA species if the ends generated by RE cleavage are intact.

2 Excessive RE in sample may negatively affect the results; contaminating endonucleases can cleave DNA at other sites, or contaminating exonucleases can damage the ends of restriction fragments. Enzyme activity and effectiveness vary among suppliers and/or batches. Consult suppliers' literature for further information.

4. DNA fragments resolved on a minigel can be recovered by electroelution for further use (Section 9-3).
5. For use other than loading on gel, samples can be purified by SS-phenol/ chloroform extraction and ethanol precipitation (Section 20-1).

SECTION **5-5.**

Agarose Gel Electrophoresis

DESCRIPTION

Agarose gel electrophoresis is used to resolve DNA fragments on the basis of their molecular weight. Smaller fragments migrate faster than larger ones; the distance migrated on the gel varies inversely with the logarithm of the molecular weight. The size of fragments can therefore be determined by calibrating the gel, using known size standards, and comparing the distance the unknown fragment has migrated. This technique can be used to resolve complex DNAs (i.e., genomic DNA) for Southern blot analysis or to resolve simpler digests of bacteriophage and plasmid clones for RE site mapping and blotting.

TIME REQUIRED

Day 1—1.5 hr
Day 2—5 hr

SPECIAL EQUIPMENT

Horizontal gel (submarine) electrophoresis apparatus for 20 × 20 cm gels (many are available, including BRL model H4)
Power supply
Microwave oven
UV-transparent plastic gel formers
UV transilluminator

REAGENTS

Agarose (Biorad Ultrapure or Seakem)
1× TAE buffer

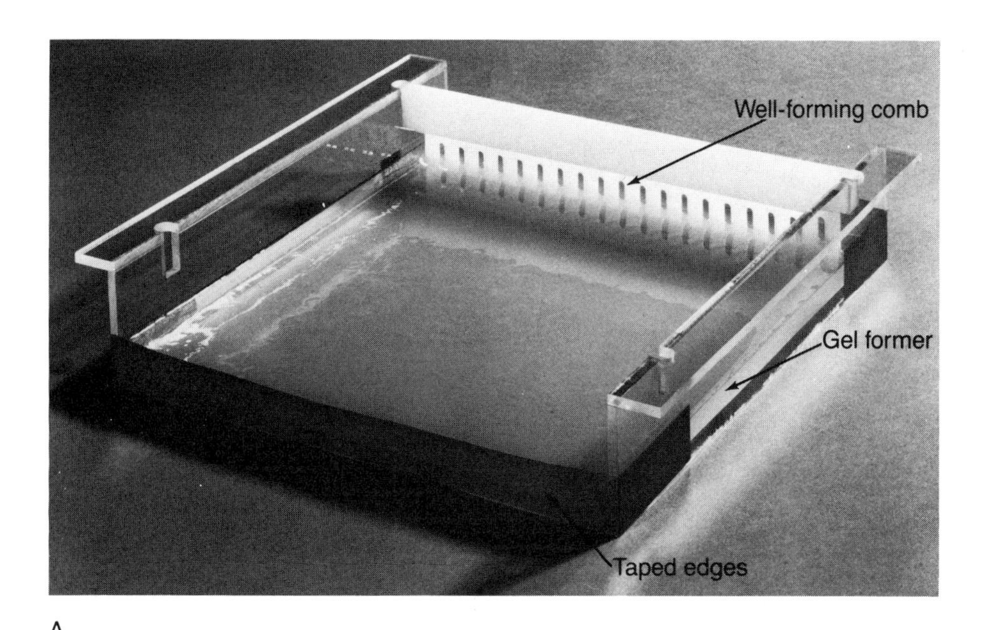

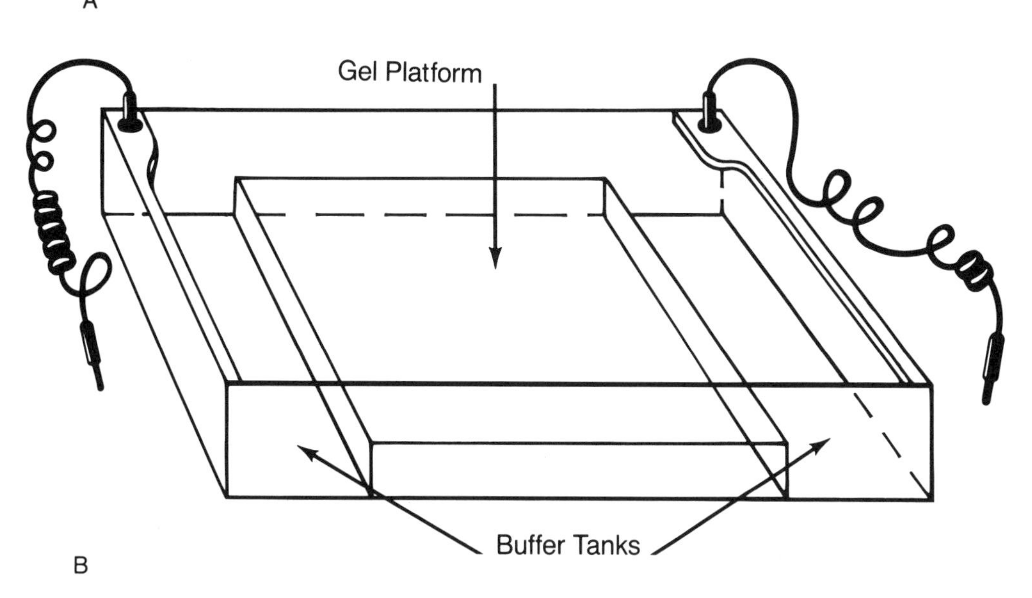

Figure 5.2
(A) Photograph of agarose gel, showing taped edges and well-forming comb in position. (B) Diagram of gel electrophoresis apparatus.

Ethidium bromide, 10 mg/ml
Gel loading buffer

METHODS

In Advance

Prepare DNA to be analyzed, including RE digest.

To Make Gel

1. Make 0.8% agarose[1] by adding 2.4 g of agarose to 300 ml of 1× TAE buffer. Agarose is solubilized by heating in a microwave oven or a boiling water bath. Allow agarose to cool to 60°C before pouring gel.
2. Commercial plastic gel formers are available for casting the gel and are recommended. If 20 × 20 cm glass plates are used, put tape around edges of clean glass plate to form a "wall." Seal tape on edges with thumbnail. Packing or autoclave tape works well.[2] With Pasteur pipette, place a line of 60°C agarose on tape perimeter to form a seal. This requires approximately 12.5 ml of the agarose solution.
3. Add 30 μl of ethidium bromide to the remaining agarose.[3] Swirl to mix, but avoid bubbles.
4. Pour remaining agarose solution into taped plate or gel former mold to make gel.
5. Place well-forming comb near one edge of gel. Clip comb to bridge above gel and make sure fingers of comb are slightly above plate, but not touching it.
6. Let gel harden until it becomes milky and opaque (approximately 20 min). Make 2 liters of 1× TAE buffer for electrophoresis tank. Pour buffer into clean gel tank. Gently remove comb and tape.
7. Place gel slab horizontally in electrophoresis tank.
8. Remove bubbles from wells with a pipette. Gels should be totally submerged in buffer, but not covered by more than 1 cm.
9. Apply 5 μl of gel loading buffer to each 25-μl sample and add samples carefully to individual wells.[4]

---

[1] 0.8% agarose gel is good for DNA pieces that range in size from 1 to 25 kb. For smaller pieces of DNA (200–2,000 bp), use 1.5% agarose; for larger pieces (20–100 kb), use 0.5%.

[2] Commercially available gel formers, with two walls positioned to hold well-forming combs, are now commonly used.

[3] Ethidium bromide is a carcinogen. Handle carefully and wear gloves.

[4] Sample is typically obtained after RE digestion of DNA (Section 5-4). Approximately 10-15 μg of genomic DNA, or 20 ng per single band of cloned DNA, is readily detected with ethidium bromide staining. If too much DNA is loaded (e.g., more than 1 μg per band) the band will be distorted. The DNA concentration can be determined by O.D.

10. Turn power on at 35–40 V. Let run overnight. DNA moves from minus to plus poles. Alternatively, the gel can be run for a shorter time at higher voltage, generally up to 100 V, unless genomic DNA is being used.

Next Day

11. The next morning, turn up power to 50 V and stop when first dye front is approximately 80% down (about 4 hr more).[5]

12. Photograph ethidium bromide–stained bands in gel using a UV transilluminator (Section 20-4) Gel is ready for Southern blotting or fragment preparation.

---

measurement (Section 20-3). To load the sample carefully, make sure there are no air bubbles in the pipette tip and do not force out the full volume (i.e., to the last partial drop) in order to prevent any air from disrupting the samples in the wells. A *Hin*dIII digest of λ DNA is useful for molecular weight standards for size calibration (see Section 9-2, step 3).

5 Gel electrophoresis can also be continued at 35-40 V for about 8 hr, if preferred.

SECTION
5-6. Southern Blot

DESCRIPTION

Southern blotting is a technique used to transfer DNA from its position in an agarose gel to a nitrocellulose (NC) filter placed directly above the gel. The DNA is denatured, neutralized, and transferred in a high-salt buffer by capillary action. The denatured, single-stranded DNA binds to the filter, is permanently bonded by baking the filter, and is later hybridized to a radiolabeled probe to detect hybridizing DNA species.

TIME REQUIRED

Day 1—3 hr
Day 2—3 hr

REAGENTS

Solution A: To make 1 liter

- 300 ml *5 M NaCl*
- 50 ml *10 M NaOH*
- 650 ml H_2O

Solution B: To make 2 liters

- 200 ml *10 M ammonium acetate*[1]
- 4 ml *10 M NaOH*
- 1,796 ml H_2O

NC filters,[2] 20 × 20 cm

---

1 Acetate is now used in buffer instead of NaCl for better removal in vacuum oven due to its higher volatility.

2 Gene Screen Plus (NEN) or nylon can be used in place of NC because they are more durable for reuse. Method conditions will change; follow manufacturers' recommendations.

METHODS

In Advance

Run electrophoretic separation of DNA on an agarose gel (Section 5-5). Take photograph of ethidium bromide–stained gel (Section 20-4)

Day 1

1. Put each agarose (typically 20 × 20 cm) gel in a plastic box with 500 ml of solution A at room temperature. Agitate periodically by hand. After 30 min, replace with 500 ml of fresh solution A. Continue periodic shaking for another 30 min. This step is designed to denature double-stranded DNA.
2. To return gel to a more neutral pH, replace liquid with 500 ml of solution B. Shake periodically, as above, for 30 min. Replace with 500 ml of fresh solution B and shake periodically for 30 min.
3. Put 500 ml of fresh solution B in another plastic dish. Wet Whatman 3MM filter with solution B to act eventually as a wick. We use a 50- to 60-cm-long piece torn from a 23-cm-wide roll. Place a 20 × 30-cm plastic or glass plate across the top of the dish and place wick over plate with ends hanging down into the solution. Smooth wick on top of plate by rolling with a 5-ml glass pipette to remove trapped bubbles that distort transfer (see Figure 5-3A).
4. Carefully lift gel from dish containing solution B in step 2 by sliding a glass plate underneath it for transport. Cover gel with a second plate. Carry gel over to wick. Flip gel over, and remove the top glass plate, and slide gel onto wick. Align gel with edge of wick and plate. Smooth gel and remove trapped bubbles by rolling a glass pipette over surface (see Figure 5-3B). Note: Gel is now upside-down, because it is best to blot from the bottom for sharper resolution and to maintain the left-to-right sample orientation on the blotted NC filters.
5. Prewet 20 × 20 cm NC filter in water.[3] Place filter directly over gel. Smooth as above.[4]
6. Prewet Whatman 3MM 20 × 20 cm filter in water or solution B.[3] Place over NC filter and smooth as above.[4] Cover with a second 3MM filter. Smooth as above.
7. Six sheets of dry, flat paper towels are placed individually over the 3MM filters.[5] They act to start capillary action, drawing liquid up through filters.
8. Place four additional 3-cm-thick stacks of folded paper towels on top (two wads in each of two directions).

---

[3] For uniformity, reduced manipulation, and decreased waste, it is helpful to buy gel formers, paper, and NC filter in a uniform 20 × 20 cm size.

[4] All smoothing is done by gently rolling a 5-ml glass pipette over the surface.

[5] We use opened Scott 175 (brown) single-fold paper towels because they have good absorbance, are inexpensive, and are easy to handle.

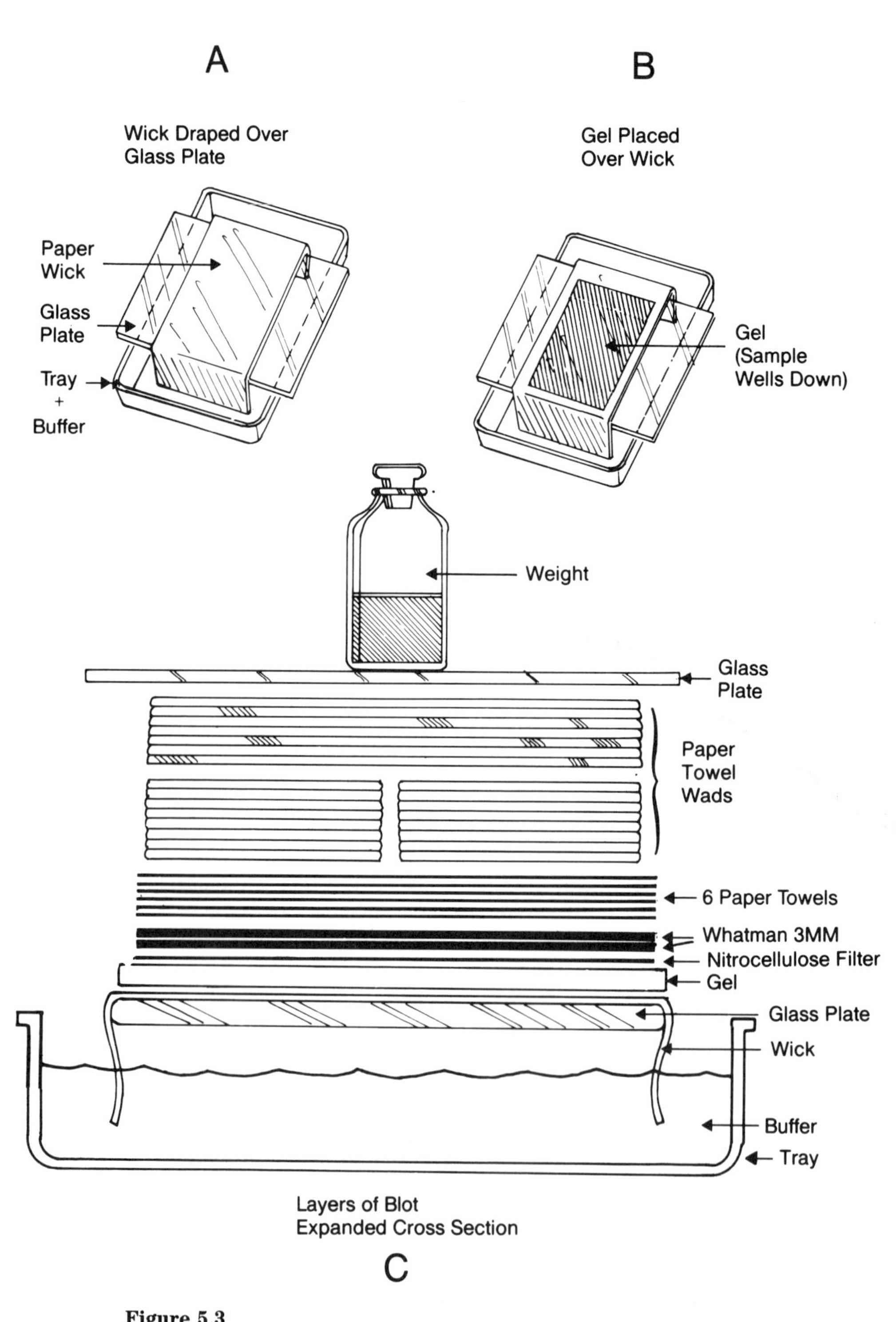

Figure 5.3
Preparation of a Southern or Northern blot. (A) Position of wick over glass plate. (B) Gel is placed on wick. (C) Schematic diagram of final layered organization of materials.

9. Top with glass plate and place a light weight (approximately 250 g) on top of plate (see Figure 5-3C). Allow 4 h to overnight for DNA transfer to NC.

Day 2

10. Remove weight, top glass, and paper towels. Slice off protruding wick ends.[6] Wick ends will easily tear off along edge of bottom plate.
11. Flip over filters/NC/gel "sandwich." Remove wick, which is now on top. Make single pinholes in centers of all sample wells down through NC to mark their position. Make additional pinholes over first well for later orientation/identification. Note that the left-to-right orientation is now in the same order as when samples were loaded.
12. Remove gel and discard. Allow a few minutes for filter to dry. Carefully mark gel number on top of NC with a ballpoint pen. It is useful to number all gels sequentially (Skilcraft ballpoint pen) and keep track of the number, date, conditions, samples, and so on in a log book. Put NC filter between two 20 × 20 cm Whatman 3MM filters. Secure on two sides with small pieces of masking tape.
13. Place in vacuum oven for 2 hr at 80°C to bake DNA onto the filter.
14. Store carefully (between 3MM filters) at room temperature until used for hybridization.

REFERENCES

Smith, G., and Summers, M., Anal. Biochem. *109:*123, 1980.

Southern, E., J. Mol. Biol. *98:*503, 1975.

---

[6] Wear gloves for all day 2 procedures.

SECTION

6

Probing Nucleic Acids with Labeled Synthetic Probes

SECTION **6-1.**

Making Synthetic DNA Probes: General Description

DESCRIPTION

Synthetic oligonucleotide DNA probes, generally 20–50 bases in length, have many uses in probing and identifying DNA or RNA samples. Two types of probes can be synthesized:

1. If the sequence of the DNA or RNA of interest is known, an exact complementary probe can be synthesized.
2. If the starting point for determining the sequence of the probe is from the amino acid sequence of a protein product, it is not possible to synthesize a single probe of significant length with complementary sequence to the RNA that uniquely codes for that protein, due to the degeneracy of the genetic code. However, understanding the genetic code and codon usage frequencies allows one to make either an educated guess, based on usage, or a "family" of probes covering all possibilities.

With the advent of automated synthesizers, it has become easier to make synthetic probes and this method has become more popular. Thus, almost all probes are now machine made, and their chemical synthesis will not be described here. Also, custom synthesis of probes is now readily available from commercial sources.

Selection of Probe

In general, the longer the probe, the stronger and more specific the hydridization will be. However, automated synthesizers often have a maximum optimal probe length (i.e., longer lengths will have a decreased yield) of approximately 50 bases.

If the RNA or DNA sequence is known, select a unique region and run a computer match (information services are available) to determine a region that is unique to that sequence (preferably over 20 bases). Determine the complementary base sequence for the region of interest (e.g., G with C or A with T). The sequence is usually written 5′→3′ (left to right), and the selected sequence will be 3′→5′ (left to right), but is conventionally rewritten in the reverse order (i.e., 5′→3′). For example:

selected sequence	5′ G C C T A A C 3′
complementary sequence for probe	3′ C G G A T T G 5′
complementary sequence reversed	5′ G T T A G G C 3′

If the RNA or DNA sequence is not known, but the amino acid sequence has been determined, a family of synthetic probes can be made based on the degeneracy of the codons (see the table below). It is likely that one of the sequences will be correct, especially if cDNA or RNA is being probed. If genomic DNA is being probed, it is possible that an intron region could interrupt the sequence and no sequence in the probe family would be correct. Typically, 16 or 32 sequences made simultaneously is the maximum number used in any one family of probes. Thus, if there are 128 possibilities, four families of 32 would be synthesized.

For example, selection of a probe for the amino acid sequence His-Val-Met could be encoded in one of eight different RNA sequences:

His Val Met

1. 5′ C A C G U U A U G 3′
2. C A C G U C A U G
3. C A C G U G A U G
4. C A C G U A A U G
5. C A U G U U A U G
6. C A U G U C A U G
7. C A U G U G A U G
8. C A U G U A A U G

To identify this sequence, eight different complementary sequences would have to be synthesized:

1. 5′ C A T A A C G T G 3′
2. C A T G A C G T G
3. C A T C A C G T G
4. C A T T A C G T G
5. C A T A A C A T G
6. C A T G A C A T G
7. C A T C A C A T G
8. C A T T A C A T G

Table 6.1 The Genetic Code.

First Nucleotide	Second Nucleotide: U		C		A		G	
U	UUU	Phe	UCU	Ser	UAU	Tyr	UGU	Cys
	UUC	Phe	UCC	Ser	UAC	Tyr	UGC	Cys
	UUA	Leu	UCA	Ser	UAA	Term.	UGA	Term.
	UUG	Leu	UCG	Ser	UAG	Term.	UGG	Trp
C	CUU	Leu	CCU	Pro	CAU	His	CGU	Arg
	CUC	Leu	CCC	Pro	CAC	His	CGC	Arg
	CUA	Leu	CCA	Pro	CAA	Gln	CGA	Arg
	CUG	Leu	CCG	Pro	CAG	Gln	CGG	Arg
A	AUU	Ile	ACU	Thr	AAU	Asn	AGU	Ser
	AUC	Ile	ACC	Thr	AAC	Asn	AGC	Ser
	AUA	Ile	ACA	Thr	AAA	Lys	AGA	Arg
	AUG	Met	ACG	Thr	AAG	Lys	AGG	Arg
G	GUU	Val	GCU	Ala	GAU	Asp	GGU	Gly
	GUC	Val	GCC	Ala	GAC	Asp	GGC	Gly
	GUA	Val	GCA	Ala	GAA	Glu	GGA	Gly
	GUG	Val	GCG	Ala	GAG	Glu	GGG	Gly

Key:

Amino acid	*Symbol*	*Amino acid*	*Symbol*
Alanine	Ala	Isoleucine	Ile
Arginine	Arg	Leucine	Leu
Asparagine	Asn	Lysine	Lys
Aspartic acid	Asp	Methionine	Met
Asn and/or Asp	Asx	Phenylalanine	Phe
Cysteine	Cys	Proline	Pro
Glutamine	Gln	Serine	Ser
Glutamic acid	Glu	Threonine	Thr
Gln and/or Glu	Glx	Tryptophan	Trp
Glycine	Gly	Tyrosine	Tyr
Histidine	His	Valine	Val

Amino acids encoded by these RNA triplet codons, read 5′ to 3′, are presented. The underlined codons UAA, UAG, and UGA are termination sequences. The codon **AUG,** in bold print, is the initiation sequence.

Sequences containing Met or Trp, with no degeneracy, are preferred, and sequences with Leu, Ser, or Arg (six possibilities each) are to be avoided.

An alternative strategy to synthesizing a family of probes is to select a single, longer (25-40 base) oligonucleotide that represents a best guess of the correct nucleotide sequence. The sequence is predicted by statistical codon usage frequency. The probe length adds codon stability, even with some mismatch.

Recently, a novel approach to synthesizing oligonucleotide probes, employing deoxyinosine residues instead of NTPs, has been described (Ohtsuka et al. 1985). The deoxyinosine residues in the resulting probe form a more stable hybrid with the DNA of interest than a mismatched base, minimizing the need to create an entire family of probes.

REFERENCES

Lathe, R., J. Mol. Biol. *183:*1, 1985.

Ohtsuka, E., Matsuki, S., Ikehara, M., Yoosuke, T., and Matsubara, K., J. Biol. Chem. *260:*2605, 1985.

SECTION **6-2.**

End Labeling of Synthetic Probes

DESCRIPTION

The enzyme T4 polynucleotide kinase will specifically transfer the γ-^{32}P label from ATP to a 5′ OH group of DNA or RNA. This reaction can be used to end label DNA and RNA molecules, such as oligonucleotides, restriction fragments, and mRNAs. The species to be labeled must have a 5′ OH group. Synthetic probes are typically supplied with a 5′ OH group. DNA restriction fragments and RNA may have a 5′ phosphate terminus, requiring dephosphorylation with a phosphatase enzyme before labeling can occur.

A method for labeling on the 3′ end using the enzyme terminal deoxynucleotidyl transferase is also presented.

TIME REQUIRED

2 hr

SPECIAL EQUIPMENT

Liquid scintillation counter

REAGENTS

Column wash buffer: To make 100 ml:

- 98 ml of *TE buffer*
- 1 ml of *5 M NaCl*
- 1 ml of *10% SDS*

G-50 Sephadex (P-L) column

10× kinase buffer:

500 mM Tris, pH 7.4
100 mM $MgCl_2$
50 mM DTT
10 mM spermidine (optional)

T4 polynucleotide kinase, 10 U/μl, in glycerol, stored at −20°C (BRL or P-L)
γ-^{32}P-ATP (3,000 Ci/mmol; 10 μCi/μl) (NEN)
10% SDS

METHODS

In Advance

Prepare G-50 Sephadex. Place in H_2O overnight to swell the beads gently. Autoclave to sterilize.[1] Alternatively, Nensorb column can be used to separate the probe from unincorporated ^{32}P-ATP (Section 10-3).

Kinase Reaction to Label Probe

1. Around inside walls of 1.5-ml microfuge tube, add individual drops (i.e., drops not touching each other) of the following:[2,3]

   1 μl oligonucleotide or DNA fragment (0.5 μg)
   2.5 μl 10× kinase buffer
   1.5 μl T4 polynucleotide kinase

2. Behind a Plexiglas shield, add 20 μl of γ ^{32}P-ATP (200 μCi) to tube. Mix by spinning drops down to the bottom of microcentrifuge for 15 sec.
3. Incubate for 45 min at 37°C. During this time, pour prepared Sephadex into 5-ml columns to fill. Wash with several volumes of column wash buffer and run column until 100 μl is left on top. Stop flow in column.

---

1 Sterilize water and microfuge tubes by autoclaving. Sephadex stock is stored at room temperature. Kontes #420160 columns with associated funnel, stopcock, and disc work well.

2 Handle radioactive reagents and waste with care. Wear gloves and use proper shielding. Properly dispose of buffers, tubes, and columns that have contained radioactive reagents.

3 Alternatively, the probe can be labeled on the 3′ end with terminal deoxynucleotidyl transferase (TdT). Dry 20 μl of ^{125}I-, ^{3}H, α-^{35}S- or α-^{32}P-dCTP in a microfuge tube in a vacuum centrifuge. Add 5 μl of 5× tailing buffer (5 mM $CoCl_2$, 1 M potassium cacodylate, 100 mM Tris-Cl, 5 mM DTT, pH 7.5) and 1 μg synthetic oligonucleotide probe in 20 μl of H_2O. After adding 20 U of TdT, incubate at 37°C for 45–120 min and continue as described in step 3. As multiple nucleotides are added to the 3′ end, their number can be monitored by running on a denaturing acrylamide gel (Section 9-4).

4. Transfer sample tube to 65°C bath for 5 min to inactivate the enzyme.
5. Spin tube again for 5 sec. Sample is now ready for removal of unincorporated ^{32}P-ATP from labeled sample by size separation on a Sephadex column (see below) or by the Nensorb method (Section 10-3).
6. Add 50 μl of column wash buffer to sample. Load sample on G-50 Sephadex column and start column flow. Sequentially add three separate 100-μl aliquots of column wash buffer to column.
7. Pour 10 ml of column wash buffer into column. Collect eluate in 15 microfuge tubes. Fill first tube and then collect 0.5 ml (approximately 12 drops) in each of the remaining 14 tubes. Stop column flow when done.[4]
8. Count 5 μl from each tube in 5 ml of scintillation fluid in a scintillation counter. Pool tubes in the leading peak with high counts.[5] Calculate the counts per minute per microliter in the pooled sample. The probe is now ready for hybridization. Store probe frozen at −20°C until used for Southern blot, Northern blot, or plaque lifts.

REFERENCE

Richardson, C. C.: In: *Procedures in Nucleic Acid Research,* Vol. 2 (G. L. Cantoni and D. R. Davies, eds.). Harper & Row, New York, 1971, p. 815.

---

[4] Close off column and dispose properly in radioactive waste.

[5] Generally two to three tubes in the first peak eluted will each contain over 10,000 cpm/μl and will be pooled. If only a small amount of labeled probe is needed, use only one or two tubes with the highest counts.

SECTION **6-3.**

Hybridization with Synthetic ^{32}P End-Labeled Probe

DESCRIPTION

This method identifies homologous DNA or RNA species bound to a filter by hybridization with a labeled synthetic oligonucleotide probe. This method may be used on circular filters to visualize plaques or colonies that contain complementary nucleotide sequences or on filters blotted from gels to visualize DNA or RNA species. Most often these filters are NC or nylon membrane equivalent (e.g., Gene Screen Plus). The base pairing between synthetic probe and DNA or RNA is governed by the complementary hydrogen bonding of nucleic acids (i.e., G with C and A with T or U). The hydrogen bonding between the pairs involves association and dissociation kinetics; increasing lengths of complementary nucleic acids provide increased stability and thus favor association. However, many other factors have been shown to influence this equilibrium. The primary influences are concentration, temperature, and salt concentration in the hybridization buffer. Thus, increased temperature or decreased salt concentration favor dissociation of the double-stranded species. Also influencing the stability of the double-stranded hybridized form are the divergence of the two sequences (mismatch), the percent of G-C bonds versus A-T bonds (a G-C bond involves three hydrogen bonds as opposed to two in A-T bonds), and the amount of formamide in the buffer. The effects obtained by varying each of these factors have been studied in model systems (see References), leading to a general formula (see step 3) that can be used to estimate the melting temperature (T_m). This is the temperature at which 50% of the double-stranded hybridized species have dissociated. Thus, the use of oligonucleotide probes to identify species with homologous complementary sequences requires an awareness of these empirically derived factors. The general formula can be used to estimate the incubation temperature (T_i) to be used (i.e., about 15°C below the T_m) and the salt and formamide concentrations in the buffer. This is only a starting point, and the exact hybridization temperature for an optimal signal-to-noise ratio for any synthetic probe usually has to be empirically determined. Under these conditions, there should be sufficient sensitivity to detect specific nucleotide sequences without interference from nonspecific hybridization to background.

TIME REQUIRED

Day 1—4 hr to set up hybridization
Day 2—3 hr to prepare hybridized filter
Day 3—1 hr to examine autoradiograph

SPECIAL EQUIPMENT

Bag-sealing apparatus
X-ray film-developing system

REAGENTS

Hybridization buffer-S
2× SSC buffer

METHODS

In Advance

Prepare ^{32}P end-labeled synthetic DNA probe by T4 polynucleotide kinase end labeling with γ-^{32}P-ATP or with terminal deoxynucleotidyl transferase with α-^{32}P-dCTP (Section 6-2).[1] Have NC filters ready for hybridization.[2]

Prehybridization

1. Place NC filter(s) in a heat-sealable plastic bag and add 10 ml of hybridization buffer to the bag. Fold and clip end of bag to seal. Incubate bag at room temperature for 2–3 hr.

Adding Probe

2. Add about 20 million cpm of probe to an additional 10 ml of buffer and add to bag. Massage buffer in bag to mix. Seal bag (Section 20-2).

---

[1] Handle radioactive samples and waste with proper care. Wear gloves throughout procedure.

[2] Filters can be round (from colony lifting) or rectangular (from gel). For NC filters the DNA or RNA needs to be bound by drying in a vacuum oven. A few filter squares can be hybridized at the same time in the same bag; up to 20 NC or 6 Gene Screen circles can be hybridized in one bag.

3. Estimate the incubation temperature (T_i) from the following formulas or from the table below:
   a. $T_i = T_m - 15°C$
   b. $T_m = 16.6 \log[M] + 0.41 [P_{gc}] + 81.5 - P_m - B/L - 0.65[P_f]$
   Where:

   M is the molar concentration of Na^+, to a maximum of 0.5 (1× SSC contains 0.165 M Na^+)

   P_{gc} is the percent of G or C bases in the oligonucleotide probe (between 30 and 70)

   P_m is the percent of mismatched bases, if known (each percent of mismatch will alter the T_m by 1°C on the average)

   P_f is the percent of formamide in the buffer

   B is 675 (for synthetic probes up to 100 bases)

   L is the probe length in bases

   Thus, for a probe of 50% G/C with no mismatches against the sequences of interest and in a buffer without formamide and containing the indicated SSC concentration the following table can be used to predict the T_i to be used as a starting point:

	T_i(°C)		
Bases	2× SSC	1× SSC	0.5× SSC
50	65	60	55
40	62	57	52
30	56	51	46
20	45	40	35

   If the number of mismatches is unknown, for example when using a probe generated from one species to identify the homologous DNA or RNA in another species, the starting T_i should be decreased. If probe length is more than 40 bases, formamide can be included in the hybridization buffer to reduce the required T_i (0.65°C for each percent formamide).

4. Incubate bag overnight at the incubation temperature determined in step 3.

Next Day

5. Remove filter from bag, being careful to keep radioactive buffer inside bag. Reseal and dispose of bag.

6. Wash filter in a plastic dish with 250 ml of 2× SSC buffer, prewarmed to incubation temperature, for 5 min.[3]

[3] If the synthetic probe is homologous to the hybridizing species and longer than 35 bases, more stringent wash conditions should be used (e.g., 1× SSC or 0.5× SSC).

7. Change to fresh buffer at 10°C lower than the incubation temperature. Incubate at this new temperature for 30 min. Change buffer two more times at 30-min intervals for a total of four washes.
8. Blot filter between two pieces of Whatman 3MM filter paper and air-dry for 30 min. Mount on 3MM filter and cover with clear plastic wrap. Mark a small adhesive label with radioactive ink and place on wrap for orientation.
9. Autoradiograph overnight at −70°C with screen and X-ray film (Section 20-5).

Next Day

10. Develop X-ray film to determine positive hybridization. Store wrapped filter in a safe place.[4,5]

REFERENCES

Anshelevich, V. V., Vologodskii, A. V., Lukashin, A. V., and Frank-Kamenetskii, V., Biopolymers *23:*39, 1984.

Bonner, T. I., Brenner, D. J., Nenfeld, B. R., and Britten, R. J., J. Mol. Biol. *81:*123, 1973.

Britten, R. J., Graham, D. E., and Nenfeld, B. R., Meth. Enzymol. *29:*363, 1974.

Cantor, C. R., and Schimmel, P. R., *Biophysical Chemistry,* W. H. Freeman, San Francisco, 1980, p. 1109.

Lathe, R., J. Mol. Biol. *183:*1, 1985.

---

4 Filters can be reautoradiographed, if necessary. If background is too high, wash filters at a higher temperature or in a more dilute SSC buffer (steps 6 and 7).

5 Filters can be rehybridized two to three times with another probe. Radioactivity from the previous hybridization can be washed off in H_2O containing 1% glycerol at 80°C for 2 min, followed by a few washes in H_2O at room temperature. The filter is now ready for rehybridization.

SECTION

Probing Nucleic Acids with Plasmid-Derived Probes

SECTION **7-1.**

Nick Translation

DESCRIPTION

The nick translation reaction is used to introduce radioactive nucleotide phosphates into unlabeled DNA for the purpose of making a probe. The reaction depends on the ability of the enzyme DNA polymerase I to initiate DNA synthesis at free 3′ OH groups, which are exposed as nicks in the unlabeled DNA. The nicks are generated in random locations by a limited digest of the DNA to be labeled with DNase I. The polymerase synthesizes new DNA in a 5′ to 3′ direction, using labeled triphosphates. The newly polymerized DNA is therefore radioactive.

TIME REQUIRED

2 hr after purification of DNA probe

REAGENTS

DNase I (DPFF grade) (Cooper Biomedical) resuspended in 5 mM $CaCl_2$ and 1 mM $MgCl_2$ in a 1 mg/ml concentration. 5-μl aliquot stored at −20°C; each aliquot is used only once

Activation buffer

10 mM Tris, pH 7.4

5 mM $MgCl_2$

BSA, Pentax fraction V, 1 mg/ml

Store at −20°C

DNA polymerase I, 5 U/μl (BM) (store at −20°C)

3 M sodium acetate, pH 7.4

SS-phenol

Nick translation buffer

0.5 M Tris, pH 7.4

0.1 M $MgSO_4$

1 mM DTT

500 μg/ml BSA, Pentax fraction V

Store in 0.5-ml aliquots at −20°C

Yeast tRNA, 10 mg/ml

Unlabeled triphosphate mix

200 μM dGTP, dATP, dTTP in 50 mM Tris, pH 7.4 (store at −20°C)

α-^{32}P-dCTP (NEN or Amersham)

Chloroform

10% trichloroacetic acid with 0.1 M sodium phosphate

95% ethanol

METHODS

In Advance

Prepare unlabeled DNA to be used in nick translation; typically, a cloned DNA probe is excised from a vector sequence with an RE (Section 8-2). The fragment is purified on an agarose gel (Section 9-1 or 9-2) or a 5% polyacrylamide gel (Section 9-4) and electroeluted (Section 9-3). For further purifying DNA fragments from gel material, it is advisable to use an Elutip minicolumn or a Nensorb column (Section 10-3).[1] Alternatively, a probe can be made from a cloned fragment nick translated with accompanying plasmid or phage vector sequences.

Nick Translation

1. Add the following reactants in a 1.5-ml microfuge tube:[2]
   a. The prepared DNA probe (0.5 μg) plus H_2O to total 7 μl.
   b. 1 μl DNase I, diluted to 1 in 10,000 parts activation buffer (two successive 1:100 dilutions of 2 μl in 200 μl). Dilute immediately before using.
   c. 4 μl unlabeled triphosphate mix (200 μM dGTP, dATP, dTTP; final concentration in reaction is 32 μM for each nucleotide).
   d. 2.5 μl nick translation buffer.
   e. 0.5 μl DNA polymerase I.

---

[1] Purification of the probe is important to remove agarose, which can affect nick translation reactions.

[2] These solutions can be added as drops on inner wall of microfuge tube. Add radioactive drop "f" last. To mix together, spin drops together in microcentrifuge.

f. 10 μl of (α-^{32}P-)dCTP (final concentration in reaction is 1–2 μM).[3]

The total incubation mixture is 25 μl.[4]

2. Incubate at 14°C for 20 min.[5]

To Purify Labeled Probe

3. Move tube to ice and add the following to the reaction:

20 μl 3 M sodium acetate
150 μl H_2O
200 μl SS-phenol
200 μl chloroform

4. Close tube and mix reactants together by vortexing. Spin for 1 min in microcentrifuge to separate phases.

5. Hold tube at an angle and transfer upper aqueous layer to a new microfuge tube containing 2.5 μl of tRNA stock solution. Discard lower organic phase and old tube (which is radioactive).

6. Add 500 μl of ethanol. Mix by vortexing. Chill tube for 30 min on ice. Spin in microcentrifuge for 10 min at 4°C. Discard supernatant (radioactive) and save precipitate in tube.

7. Add 1 ml of H_2O to dissolve precipitate. Mix by vortexing.

8. To determine the activity of the labeled probe, place 2 μl of sample in a new tube and fill tube with 10% trichloroacetic acid (TCA). Pour this mixture through a glass fiber (e.g., Whatman GF/C) or NC filter under vacuum. Fill tube again with TCA and pour through filter. Fill tube with 95% ethanol and pour through filter. Dry filter with a heat lamp and count in liquid scintillation counter. Estimate the specific activity of the probe. A 2-μl aliquot of a 1-ml resuspension of a 0.5-μg DNA sample represents 1 ng of labeled DNA. Specific activity should be $1–5 \times 10^5$ cpm per nanogram of DNA. In a successful nick translation, 20–40% of the label will be incorporated into the probe.

---

[3] Use appropriate precautions for safe handling of radioactive reactants and waste. To make a probe with higher specific activity, use more than one labeled dNTP. Adjust the unlabeled mixture accordingly by deleting the same dNTP from the unlabeled mix (step 1c).

[4] The reaction volumes in step 1 can be doubled if more probe is needed for large screening procedures.

[5] To optimize the nick translation system, it is advisable to calibrate the time needed to obtain optimal incorporation of ^{32}P into DNA. One way to accomplish this is to perform a time course in which the incorporation of ^{32}P into 0.5 μg of DNA of interest is monitored as a function of time. Typically, optimal incorporation (20–40% incorporation of free label) is achieved within 20–40 min.

9. Store sample at 4°C or frozen at −20°C in a plastic shielded box. The sample will last approximately 1–2 weeks. It is now ready to use for hybridization.
10. Calculate how many microliters of sample are needed to give about 5×10^7 cpm. This amount will be used for hybridization (see next section).

REFERENCE

Rigby, P., Dieckmann, M., Rhodes, C., and Berg, P., J. Mol. Biol. *113:*237, 1977.

SECTION

7-2. DNA Hybridization (Southern Blot Hybridization)

DESCRIPTION

This is the procedure for hybridizing a ^{32}P-labeled DNA probe generated by nick translation to fragments of DNA separated on an agarose gel and then blotted to NC. The method is identical to the one used to hybridize nick-translated probes to RNA blots (Section 11-6).

TIME REQUIRED

Day 1—1 hr
Day 2—3 hr

SPECIAL EQUIPMENT

Plastic bag sealer
Autoradiography apparatus

REAGENTS

Salmon sperm DNA stock solution, 2 mg/ml
10 M NaOH
2.0 M Tris, pH 7.4
1.0 M HCl
Hybridization buffer-N
Wash solution A: To make 2 liters:
 200 ml *20× SSC*
 1,780 ml H_2O
 20 ml *10% SDS*

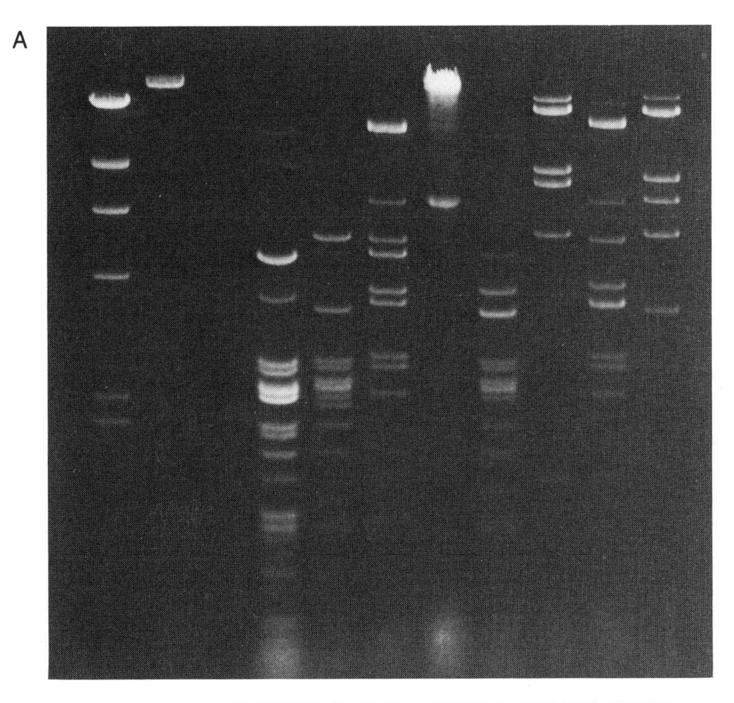

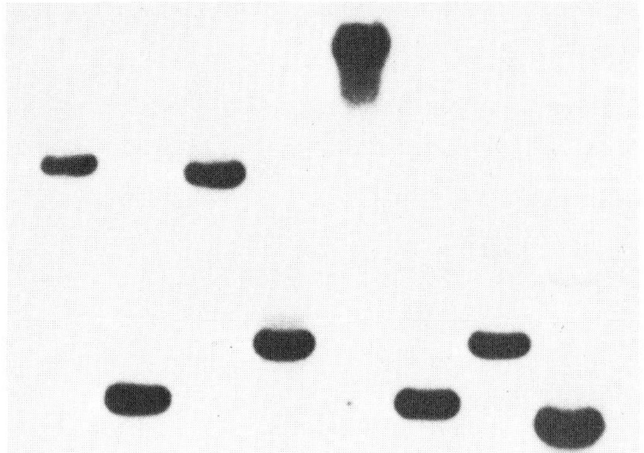

Figure 7.1
(**A**) Ethidium bromide staining of DNA in an agarose gel visualized under UV light. Plasmids from a genomic library were cut with REs and samples run on the gel. Note specific bands of cut DNA and *Hin*dIII cut λ DNA markers in left lane.
(**B**) Autoradiograph of Southern blot from similar gel. The agarose gel was blotted to NC and the blot was hybridized with an end-labeled oligonucleotide probe to identify which fragments of the genomic library are to be used for further study.

Wash solution B: To make 1 liter:

5 ml *20× SSC*
985 ml H_2O
10 ml *10% SDS*

METHODS

In Advance

Prepare ^{32}P-labeled DNA probe by nick translation (Section 7-1). Make wash solutions.

To Prehybridize Filters

1. Place 20 × 20 cm NC filter, from Southern blot, flat in heat-sealable polyethylene bag. Pour 20 ml of hybridization buffer-N over filter.[1,2] Push out any air from bag and distribute buffer evenly over filter. Let stand for at least 5 min, or until probe is prepared, at room temperature.

To Denature Double-Stranded Probe

2. Put an appropriate volume of probe sample (equivalent to 1–5 × 10^7 cpm) in a 50-ml plastic screw-cap tube.

3. Add in the following order:
a. 500 μl salmon sperm DNA stock.
b. 50 μl of 10 M NaOH. Add directly to sample.
c. 300 μl of 2 M Tris, pH 7.4.
d. 475 μl of 1 M HCl, added dropwise. Mix gently by swirling while adding.

To Hybridize

4. Add 10 ml of hybridization buffer-N to tube. Add tube contents to bag from step 1. Heat-seal bag (Section 20-2).

5. Incubate overnight submerged in a water bath at 42°C.

---

[1] Too little hybridization buffer can lead to high background. Usually, 20 ml plus probe is sufficient for a 20 × 20 cm filter. For a larger blot, add more buffer; 30 ml is a good volume to use routinely.

[2] More than one filter can be used per bag; 10 circles is the maximum recommended number.

Next Day

6. Remove filter from bag and place filter in a plastic dish. Reseal bag and place in radioactive waste. Wash filter in 500 ml of wash solution A for 5 min at room temperature. Remove buffer and replace with fresh buffer three more times at 20-min intervals.
7. Follow with two 20-min washes of filter in wash solution B at 45–65°C.[3]
8. Place filter between two pieces of Whatman 3MM paper or paper towels to blot dry. Air-dry filter for about 30 min.
9. Replace dry filter in sealable bags or place on 20 × 20 cm sheet of Whatman 3MM paper and cover with clear plastic wrap. Label filter for orientation with ^{35}S ink on a stick-on label.
10. Autoradiograph with Kodak XAR-5 film and screen at −70°C. Expose overnight and develop (Section 20-5).[4]

REFERENCE

Wahl, G., Stern, M., and Stark, G., Proc. Natl. Acad. Sci., USA *76*:3683, 1979.

---

[3] The wash temperature will determine the signal-to-noise ratio. More stringency (less background) will be gained with a higher temperature. The homology of probe to DNA, probe length, and salt concentration can also affect the signal-to-noise ratio. Adjust wash conditions for specific cases.

[4] If more or less exposure is desired, repeat autoradiography for an appropriate period. If background is too high, go back to step 7 and rewash at a higher stringency (i.e., higher wash temperature).

SECTION

Plasmid DNA Preparation

SECTION **8-1.**

Transformation of Bacteria

DESCRIPTION

This is a procedure for transforming bacterial cells with circular DNA plasmids containing an origin of autonomous DNA replication and a selectable marker (antibiotic resistance). Logarithmically growing cells are made permeable to DNA by incubation in $CaCl_2$, and the DNA is taken up from the surrounding buffer. This uptake is facilitated by a brief heat shock. Bacterial cells that have successfully taken up the DNA plasmid are then selected for growth on an agar plate containing the antibiotic. Cells not containing the plasmid will not grow on these plates, and successful tranformants will form discrete bacterial colonies.

TIME REQUIRED

Day 1—6–7 hr
Day 2—1 hr

REAGENTS

LB medium (sterile)
50 mM $CaCl_2$ (sterile)
Glycerol sterilized by autoclaving
LB/agar/antibiotic plates (use appropriate antibiotic, depending on vector) (Section 20-6)

METHODS

In Advance

Obtain bacterial strains (e.g., LE392, HB101, JM103, JM107, or JM109 cells) for pBR322-derived plasmids. Streak out LE392 or HB101 on an LB plate, or JM103,

JM107, or JM109 on a minimal agar plate, to obtain a stock plate of bacterial colonies.[1] See Section 20-6 for a description of preparing agar plates. Obtain purified circular plasmid containing insert of interest to transform cells (from preparation, purchase, or as a gift). LB medium/agar/antibiotic plates must be prepared for selective growth of transformants. Use appropriate antibiotic (Section 20-6).

Grow Bacterial Cells[2]

1. Add 40 ml of LB medium to sterile culture flasks. Sterilize flask neck in flame.[3]
2. Sterilize a metal transfer loop over open flame. Cool loop by touching on edge of agar.[4] Scrape cells (one colony) from stock plate with loop.
3. Transfer cells to medium and shake cells off loop. Put flask in shaking water bath at 37°C.
4. Grow cells to an O.D. at 600 nm of 0.3 to 0.5. This takes several hours.[5]
5. Spin cells at 2,500 × *g* for 5 min at room temperature.
6. Gently resuspend pellet of cells in 2–4 ml of 50 mM $CaCl_2$. Bring up to one-half of original volume (e.g., 20 ml) with 50 mM $CaCl_2$. Cells are now very fragile.
7. Incubate for 30 min on ice. Cells are now called *competent.*[2]
8. Centrifuge cells for 5 min at 2,500 × *g* at 4°C. Resuspend in one-tenth of the original volume (4 ml) of ice cold 50 mM $CaCl_2$. For increased competence, cells can be stored overnight at 4°C before use. Cell viability will decrease after 2 days.
9. In a sterile tube, mix about 0.1 μg of plasmid vector DNA or ligated plasmid vector with 200 μl of competent cells from step 8.
10. Incubate tube on ice for 30 min.
11. Heat shock cells by transferring tube to a 42°C water bath for 2 min. Alternatively, a 37°C bath for 5 min can be used.

---

1 Agar plates with stock bacterial cells can be stored at 4°C for about 1 month. Seal edges of dish with tape or plastic film.

2 Alternatively, frozen competent cells (e.g., HB101) can be purchased from either BRL or Vector Cloning Systems. See suppliers' protocol for methods to be used with these cells.

3 Follow sterile procedure for all steps in this method. All transfers must be kept sterile. Autoclave all tubes before using.

4 You can use a sterile plastic loop (without flaming).

5 It is very important that cells grow logarithmically to achieve optimal transfection efficiency.

12. Add 1 ml of LB medium and grow for 45 min at 37°C, with shaking or in rotary drum, to allow expression of the antibiotic resistance gene.

Plating of Cells

13. Use bent-end glass rod[6] (sterilized with 70% ethanol and flaming) to spread cells over surface of LB/agar/antibiotic plates. Spread several different amounts (e.g., 50, 100, and 200 μl) on separate plates to determine empirically the amount needed to generate adequate numbers of well-separated colonies. Store remaining cells at 4°C for further plating, if necessary.
14. Add 100 μl of untransformed bacterial cells from step 8 to another plate (with antibiotic) as a negative control.
15. Place plates in a 37°C incubator, agar side down, for 20 min to allow medium to dry.
16. Invert plates (agar on top) and grow overnight at 37°C. Colonies will become apparent at 12–16 hr.
17. Colonies are now ready for colony hybridization (Section 15-3), mini-prep DNA analysis of plasmids (Section 8-4), preparative plasmid growth (Sections 8-2 and 8-3), or storage.

Long-Term Storage of Transformed Bacteria: Glycerol Stocks

18. Remove single colonies of cells from plate with sterile transfer loop and add to individual tubes containing 2 ml of LB medium supplemented with the appropriate antibiotic. Grow, with shaking, until culture becomes cloudy.
19. Take a 0.8-ml aliquot of cells from each tube and place in small sterile tubes. Add 0.2 ml of sterile glycerol to each tube. Mix. Freeze cells at −70°C. The bacterial transformants may be recovered by streaking the glycerol stock onto an appropriate LB agar-antibiotic plate to recover colonies.

REFERENCES

Cohen, S., Chang, A. C. Y., and Hsu, L., Proc. Natl. Acad. Sci., USA *69*:2110, 1973.

Dagert, M., and Ehrlich, S., Gene *6*:23, 1979.

---

[6] Shaped similar to a flat-bottom hockey stick.

SECTION

8-2. Plasmid DNA Preparation: Triton-Lysozyme Method

DESCRIPTION

This method is used to prepare plasmid DNA (or any small circular DNA, such as M13 RF DNA) from *E. coli.* The bacteria are partially lysed with lysozyme and Triton, allowing the plasmid to escape the cell wall into the supernatant. The larger *E. coli* chromosomal DNA is trapped in the cell wall. The lysate is cleared of cell debris, and the plasmid-containing supernatant is banded in a CsCl/ ethidium bromide equilibrium gradient to purify closed circular DNA from any sheared linear plasmid or *E. coli* DNA fragments (linear and circular DNA band at different densities in this gradient). This is a commonly used method that generates highly purified plasmid DNA, free of RNA contamination, but requires a considerable amount of time and specialized ultracentrifuge equipment (vertical rotors and ultracentrifuges).

TIME REQUIRED

Day 1—2–3 hr
Day 2—1 hr
Day 3—5 hr
Day 4—1 hr
Day 5—3 hr

SPECIAL EQUIPMENT

Ultracentrifuge (with Du Pont/Sorvall TV850 and TV865 or Beckman VTi50 and VTi65 vertical rotors, caps, and wrenches)
Refractometer
Biodryer apparatus (vacuum centrifuge)
Quick-seal ultracentrifuge tubes and sealing apparatus or Du Pont/Sorvall tube crimping system

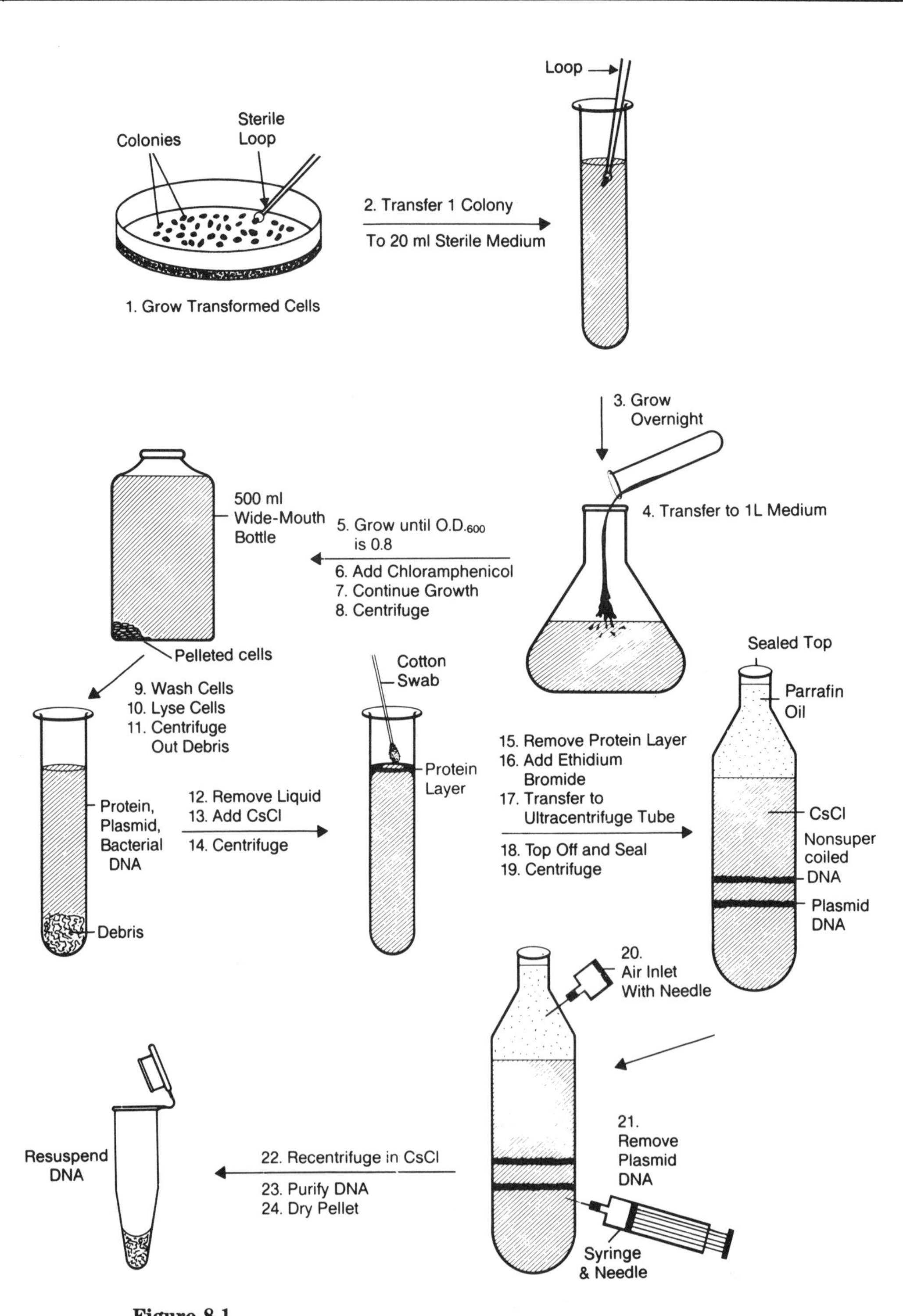

Figure 8.1
Plasmid preparation. Plasmids of interest are selected and grown on a large scale. Plasmid DNA is freed from the bacteria and isolated by two successive centrifugations through CsCl.

Quick-seal tubes (Beckman); VTi50—25 × 89 mm tube; VTi65—13 × 51 mm tube or Du Pont/Sorvall VT850 and VT865 crimp top tubes

REAGENTS

Components for M9-glucose minimal medium; see step 1 below
Chloramphenicol, 34 mg/ml, in ethanol (store at −20°C)
10% sucrose in 50 mM Tris, pH 8.0
Lysozyme solution (30 mg/ml H_2O); make fresh before using (Sigma)
1% Triton X-100
CsCl solution (mix 19 g optical-grade CsCl with 20 ml H_2O)
Ethidium bromide, 10 mg/ml
TE buffer
n-Butanol (H_2O saturated)
0.5 M EDTA, pH 8.0
3 M sodium acetate, pH 7.4
SS-phenol
Chloroform

METHODS

In Advance

Add one colony to 20 ml of LB medium supplemented with the appropriate antibiotic.[1] Grow overnight at 37°C, with shaking.

Preparing the Medium

1. Make the five components for minimum bacterial growth medium (M9-glucose).

   a. $Na_2HPO_4 \cdot 7H_2O$ 60 g
   KH_2PO_4 30 g
   NaCl 5 g
   NH_4Cl 10 g
   Dissolve in 900 ml of H_2O and adjust pH to 7.4 with 10 N NaOH.

   b. 20 g glucose; add H_2O to 100 ml. Sterilize.

   c. 100 g casein-hydrolyzed amino acids (casamino acids) adjusted to 1 liter in H_2O.

   d. *1 M $MgSO_4$*

   e. *1 M $CaCl_2$*

---

[1] Commonly used antibiotics are tetracycline or ampicillin for pBR-derived plasmids. The choice of antibiotic depends on the location of the insert within the plasmid. Insertion of DNA into these sites will inactivate the ampicillin or tetracycline gene. For ampicillin, use 50 μg/ml of growth medium; for tetracycline, use 12.5 μg/ml.

2. Autoclave each of the above, except solution b, individually and let cool.
3. Add solutions a and b together (a alone is 10× M9; a plus b is 10× M9-glucose).
4. To prepare 1 liter of minimal medium:

   a. Start with 800 ml of sterile H_2O.
   b. Add 1 ml of 1 M $MgSO_4$ and 0.1 ml of 1 M $CaCl_2$. Mix.
   c. Add 100 ml of a/b mixture from step 3.
   d. Add 100 ml of solution c.
   e. Add antibiotics, as appropriate.[2]

Inoculate Medium

5. Add 20 ml of cells previously grown overnight in LB medium with the appropriate antibiotic (see "In Advance") to 1 liter of minimal medium in a 2.5-liter firmback or 4-liter Erlenmeyer flask.
6. Grow with shaking until O.D. at 600 nm is 0.8.[3] Check O.D. every 30 min after the second hour.
7. When O.D. is 0.8, add 5 ml chloramphenicol solution to amplify copy number of plasmid while stopping bacterial DNA replication.
8. Shake at 120 rpm overnight at 37°C on an orbital shaker.

Harvesting

9. Spin cells in four 250-ml or two 500-ml wide-mouth bottles at 4,000 × *g* for 5 min at 4°C. Discard supernatant.
10. Resuspend and recombine cells in 10 ml of TE buffer and transfer to clean polycarbonate ultracentrifuge tubes (e.g., Ti 50.2 screw-cap tubes).
11. Spin at 4,000 × *g* for 5 min at 4°C.
12. Resuspend pellet in 8 ml of 10% sucrose in 50 mM Tris, pH 8.0.
13. Add 1 ml of freshly made lysozyme solution to digest peptidoglycan linkages in cell walls. Mix. Incubate on ice for 10 min.
14. Add 2 ml of 0.5 M EDTA to chelate divalent cations and inhibit DNases. Mix. Place tube on ice for 5 min.
15. Add 5 ml of 1% Triton X-100 (in H_2O). Mix. Warm to 37°C for 2 min or until lysis occurs. Lysis is apparent when viscosity increases and bacterial debris is visible.
16. Spin at 192,000 × *g* (e.g., 40,000 rpm with a Beckman Ti 50.2 or Du Pont/Sorvall rotor) for 30 min at 4°C.

---

[2] See note 1.

[3] A light-scattering measure of cell density. Place a sample of medium only in a spectrophotometer set at 600 nm to zero instrument. Read the O.D. of the sample (Section 20-3).

17. Pour supernatant (clear lysate) into a 50-ml capped tube. Bring volume to 20 ml with H_2O. Add 19 g of CsCl. Mix to dissolve CsCl.

18. Spin for 5 min at 2,000 $\times$ *g*.

19. Remove top protein layer with cotton-tipped swab and discard.

20. Check the refractive index of a 10-μl aliquot at this point with a refractometer. It should equal 1.395–1.400. Add more water or CsCl as necessary to adjust refractive index.

21. To remaining liquid, add 400 μl of ethidium bromide solution.[4]

22. Transfer to sealable centrifuge tubes. Fill the remaining volume of the tube completely with paraffin oil, seal tube, and spin in vertical rotor at 225,000 $\times$ *g* overnight at 20°C (e.g., Beckman VTi50 or Du Pont/Sorvall TV850 at 48,000 rpm). Tubes are either Beckman polyallomer heat sealable, 25 $\times$ 89 mm, or Du Pont crimp-top, 35 ml.

Next Day

23. Stop centrifuge without brake. Check for two bands with UV light. Remove top of the tube or pierce top with an 18-gauge needle. Test plunger of 2-ml syringe for easy movement. Pierce through tube side with 18-gauge needle connected to this syringe, 5 mm below lower band. Aim up and remove lower band with syringe.[5] Place collected material in smaller ultracentrifuge tube (e.g., Beckman Quickseal, 13 $\times$ 51 mm tube) and fill to the top with CsCl solution.[6] Seal tube.

24. Spin for at least 8 hr at 18°C, 278,000 $\times$ *g* (e.g., Beckman VTi65 or Sorvall TV865 rotor at 54,000 rpm).

Next Day—Plasmid Extraction

25. Stop centrifuge without brake. Visualize bands under UV light. Remove band with needle (lower band if two bands are present), exactly as described in step 23, and place in 10-ml capped polypropylene tube. Band volume should be approximately 1 ml.

26. For each milliliter of CsCl band solution, add 3 ml of TE buffer.

27. Add 4 ml of *n*-butanol. Mix. Briefly centrifuge to separate phases. Remove upper butanol layer by vacuum suction and discard. Butanol is now orange from ethidium bromide.

28. Extract three more times with 4 ml of *n*-butanol, as in step 27. The orange color will fade, but the interface is still visible.

---

[4] Ethidium bromide is a mutagen; handle it properly.

[5] The upper band is linear bacterial chromosomal and nicked circular and linear plasmid DNA. The second (saved) band is the desired closed circular plasmid DNA.

[6] At this point, 20 μl of ethidium bromide solution can be added if the solution does not appear pink.

29. Transfer lower, aqueous phase to a 30-ml Corex tube.
30. For each milliliter of original CsCl band solution (step 25), add 8 ml of ethanol. Mix. Place on ice for 20 min. Do not chill on dry ice; lower temperatures will precipitate CsCl salt.
31. Spin at 12,000 × *g* for 10 min at 4°C.
32. Discard supernatant and drain excess liquid from pellet. Resuspend pellet in 350 μl of TE buffer. Transfer to 1.5-ml microfuge tube. Rinse Corex tube with additional 100 μl of TE buffer and add to microfuge tube.
33. Add 0.5 ml of SS-phenol and 0.5 ml of chloroform. Mix. Spin in microcentrifuge for 2 min. Transfer upper aqueous layer to a new microfuge tube.
34. Add 50 μl of 3 M sodium acetate. Mix.
35. Add 1 ml of ethanol. Mix. Place on ice to precipitate for 10 min.
36. Spin in microcentrifuge for 5 min at 4°C.
37. Discard supernatant. Wash pellet in 500 μl of 80% ethanol. Mix by vortexing. Spin in microcentrifuge for 5 min at 4°C.
38. Discard supernatant and dry pellet in vacuum centrifuge.
39. Resuspend pellet in 500 μl of TE buffer.
40. Put 5 μl of sample in 1 ml H_2O. Measure O.D. at 260 nm to determine DNA concentration (Section 20-3).
41. Store sample at −20°C. The sample is very stable.

REFERENCES

Clewell, D., and Helinski, D., J. Bacteriol. *110:*1135, 1972.

Radloff, R., Bauer, W., and Vinograd, J., Proc. Natl. Acad. Sci. USA *57:*1514, 1967.

SECTION **8-3.**

Large-Scale Alkaline Lysis Method: Plasmid Purification

DESCRIPTION

This method is used for purifying plasmid DNA from *E. coli* suitable for most molecular biological applications. It has an important advantage over the Triton-Lysozyme method (Section 8-2) in that it does not require ultracentrifugation as a step in purification and requires considerably less time. The single disadvantage of this method is that a small amount of *E. coli* chromosomal DNA and RNA is present in the final product, which complicates precise quantitation of plasmid yield using O.D. measurements.

TIME REQUIRED

1 day after culture is grown

SPECIAL EQUIPMENT

1 × 25 cm column for Sepharose 4B

REAGENTS

Solution A
- 50 mM glucose
- **10 mM EDTA**
- 25 mM Tris, pH 8.0

Solution B
- Lysozyme, 20 mg/ml (make fresh for each use), dissolved in solution A

Solution C

0.2 M NaOH

1% SDS

Solution D

29.4 g potassium acetate

11.5 ml glacial acetic acid

Add H_2O to 100 ml

Sepharose 4B, preswelled in 4B buffer

4B buffer

0.5 M NaCl

50 mM Tris, pH 7.4

10 mM EDTA

3 M sodium acetate, pH 7.0

0.3 M sodium acetate, pH 7.0

RNase A, 10 mg/ml (heat to 70°C for 10 min to inactivate DNase)

SS-phenol

Chloroform

Isoamyl alcohol

Ethanol

TE buffer

Ethidium bromide, 10 mg/ml

METHODS

In Advance

Preparatively, grow a 250 ml culture of *E. coli* containing the plasmid, as described in Section 8-2, steps to 1 to 8.

Preswell Sepharose 4B in 4B buffer. Pour into a 1 × 25 cm column.

Procedure

1. Collect cells from 250 ml of culture by centrifugation in 500-ml bottles at 4000 × *g* for 5 min. Discard supernatant.
2. Resuspend cell pellet in 2 ml of solution A and transfer to a 10-ml snap-cap polypropylene tube.
3. Add 0.5 ml of solution B. Mix thoroughly. Incubate tube for 10 min on ice.
4. Add 5 ml of solution C. Mix gently. Incubate tube for 10 min on ice.
5. Add 4 ml of solution D. Swirl to mix. Let tube stand on ice for 10 min.
6. Spin down white precipitate by centrifugation at 12,000 × *g* for 10 min in either Corex or snap-cap polypropylene tubes.

7. Transfer supernatant to a new tube. Be careful not to transfer any precipitate. Discard the precipitate.
8. Add heat-inactivated RNase A to the supernatant at a final concentration of about 20 μg/ml. Incubate for 15 min at room temperature.
9. Extract with an equal volume of SS-phenol:chloroform (1:1). Separate phases by centrifugation at 12,000 × *g* for 10 min. Transfer upper aqueous phase to a new tube. Repeat this extraction one more time.
10. Transfer the upper aqueous phase to a new tube and extract with an equal volume of chloroform. Separate phases by centrifugation at 12,000 × *g* for 2 min. Transfer upper aqueous phase to a new tube.
11. Add one-tenth of 1 volume of 3 M sodium acetate, pH 7.0, and 2.5 volumes of ethanol. Precipitate nucleic acids on ice for 10 min. Collect precipitate by centrifugation at 12,000 × *g* for 10 min. Discard supernatant and resuspend pellet in 1 ml of 4B buffer. Apply the sample to a 1 × 25 cm Sepharose 4B column. Run column in 4B buffer.[1]
12. Collect 1-ml fractions. Monitor fractions to determine which contain plasmid DNA by O.D. measurement at 260 nm[1] (Section 20-3). Plasmid DNA is in the first peak emerging from the column (this peak is in the void volume, which is usually in fractions 3 to 8). Contaminating RNA may be found in later fractions, depending on the efficiency of the RNase digestion.
13. Pool all plasmid-containing fractions. Extract the pooled sample with an equal volume of SS-phenol:chloroform (1:1). Separate phases by centrifugation at 12,000 × *g* for 2 min. Transfer upper phase to a new tube.
14. Add 2 volumes of ethanol. Precipitate DNA on dry ice for 10 min. Spin down precipitate by centrifugation at 12,000 × *g* for 10 min. Decant and discard supernatant.
15. Resuspend pellet in 0.4 ml of 0.3 M sodium acetate, pH 7. Transfer to a microfuge tube. Add 0.8 ml of ethanol. Precipitate DNA on dry ice for 10 min. Collect DNA by centrifugation in a microcentrifuge for 5 min at 4°C. Decant supernatant and discard.
16. Rinse pellet in 0.5 ml of 80% ethanol. Spin for 5 min in a microfuge at 4°C. Decant supernatant and discard. Air-dry pellet. Resuspend in TE buffer. Determine approximate concentration of DNA by O.D. measurement at 260 nm (Section 20-3).[2] Store DNA at −20°C. Sample is very stable.

---

[1] The Sepharose column steps 11 and 12 may be replaced by a polyethylene glycol precipitation to separate plasmid from small ribonucleotides, as described by P. Krieg and D. Melton (see Promega Biotec catalog for description).

[2] The plasmid-containing fractions may also be identified using an ethidium bromide agarose plate. This plate is prepared by adding 0.4 g of agarose to 40 ml of TE buffer. Boil solution for 1 min. Allow solution to cool to about 60°C and add 4 μl of ethidium bromide solution. Pour the solution into a Petri dish and allow agarose to harden. Spot 5 μl of each fraction onto the plate. Allow 20 min for spots to soak into the plate. Illuminate the plate with UV light. Plasmid-containing fractions are in the first peak staining positively with ethidium bromide (usually fractions 3 to 8).

SECTION **8-4.**

Plasmid "Mini-Prep" Method

DESCRIPTION

The plasmid "mini-prep" method is useful for preparing partially purified plasmid DNA in small quantities from a number of transformants. It relies on an alkaline SDS lysis to free the plasmid DNA from the cell, leaving behind the *E. coli* chromosomal DNA with cell wall debris. After the solution is cleared of cell wall debris, the supernatant is retained. Plasmid DNA is then purified from the supernatant along with substantial quanities of *E. coli* DNA. This method provides enough partially purified plasmid DNA for a rapid analytical restriction digest analysis of plasmids (and Southern hybridization, if desired) before large-scale growth. This method can also be used when a small amount of plasmid DNA (e.g., 10 μg) rather than large amounts (e.g., hundreds of micrograms) of plasmid DNA are required.

TIME REQUIRED

2 hr

REAGENTS

LB medium

Glucose buffer

25 mM Tris, pH 8.0

50 mM glucose

10 mM EDTA

Lysozyme solution (8 mg/ml in glucose buffer); dissolve lysozyme in glucose buffer just before use

Isopropanol

0.2 M NaOH with 1% SDS (make fresh each time)
- 5 ml of *10% SDS*
- 44 ml of H_2O
- 1 ml of *10M NaOH*

Potassium acetate solution: To make 100 ml:
- 29.4 g potassium acetate
- 11.5 ml glacial acetic acid
- Add H_2O to 100 ml

SS-phenol
Chloroform
3 M sodium acetate, pH 7.4
TE buffer

METHODS

In Advance

Select a bacterial colony. Remove colony with sterile loop and place in 20 ml of LB medium (including appropriate antibiotic[1]). Grow at 37°C, with shaking, until culture is saturated.

Prepare Sample

1. Centrifuge cells for 10 min at 2,000 × *g*.
2. Resuspend pellet in 450 μl of glucose buffer.
3. Add 150 μl of lysozyme solution. Incubate for 5 min at room temperature.
4. Transfer to a 15-ml Corex centrifuge tube.
5. Add 1.2 ml of 0.2 M NaOH with 1% SDS. Mix. Place on ice for 5 min. Mix again.
6. Add 900 μl of ice cold potassium acetate solution. Mix.
7. Centrifuge for 10 min at 12,000 × *g* at 4°C.
8. Pour supernatant into new Corex tube. Be careful not to transfer any white pellet. Discard the pellet.
9. Add 1.5 ml of isopropanol. Mix by vortexing.
10. Place in −20°C freezer for 15 min.
11. Centrifuge for 15 min at 12,000 × *g* at 4°C.
12. Discard supernatant. Resuspend pellet in 400 μl of TE buffer. Transfer to 1.5 ml microfuge tube.

---

[1] Ampicillin (50 μg/ml of medium) or tetracycline (12.5 μg/ml of medium), depending on location of insert in plasmid used.

Sample Extraction

13. Add 40 μl of 3 M sodium acetate.

14. Extract DNA with SS-phenol/chloroform and ethanol precipitate, as in Section 20-1.

15. Discard ethanol. Air-dry pellet for 5 min.

16. Resuspend DNA in 50 μl of TE buffer.

17. Plasmid DNA is now ready for further use.[2]

18. Store plasmid preparation at −20°C.

REFERENCES

Birnboim, H., and Doly, S., Nucleic Acids Res. *7*:1513, 1979.

Ish-Horowicz, D., and Burke, J., Nucleic Acids Res. *9*:2989, 1981.

---

[2] Although RNA is also found in plasmid DNA isolated in this fashion, the RNA will not interfere with most further procedures such as RE digestion, subcloning, or ligation. Because of the RNA, the O.D. measurement for DNA concentration will not be accurate. To measure the concentration of plasmid DNA, electrophorese an aliquot of digested plasmid and compare the ethidium bromide staining intensity to a known DNA standard (e.g., *Hin*dIII cut λ bacteriophage DNA; see Section 9-1).

SECTION

DNA Restriction Fragment Preparation

SECTION **9-1.**

Minigels

DESCRIPTION

Minigels are useful for rapidly separating small amounts of DNA fragments using agarose gel electrophoresis. They are especially useful when only small amounts of DNA are available for analysis or when it is desirable to monitor rapidly the effectiveness of various enzyme reactions between sequential steps in a procedure.

TIME REQUIRED

2 hr

SPECIAL EQUIPMENT

Minigel horizontal electrophoresis apparatus
Power supply (available from many suppliers)
UV transilluminator

REAGENTS

1× TAE buffer
Agarose stock
 300 ml of 1× TAE buffer with 0.8% agarose (2.4 g)
 Microwave to bring into solution before use

Ethidium bromide, 10 mg/ml
Gel loading solution

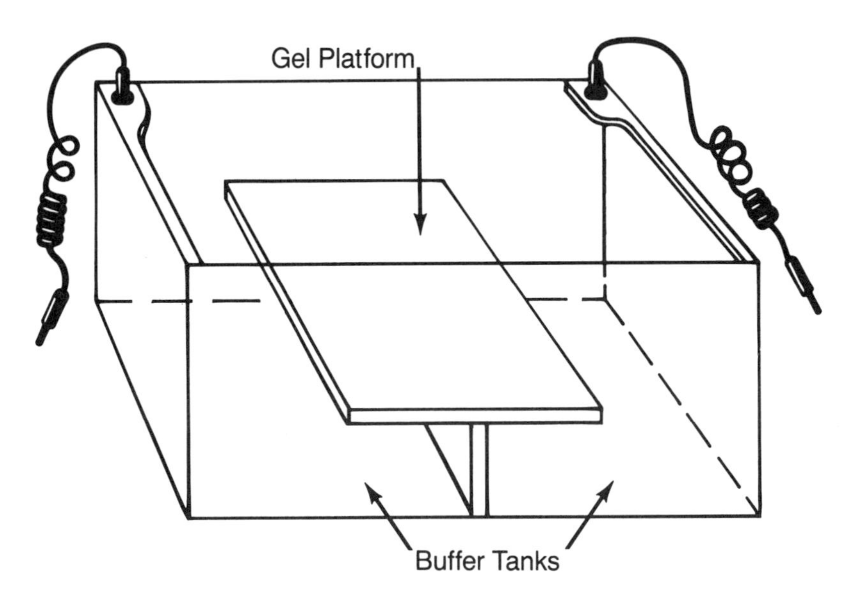

Figure 9.1
Diagram of a Minigel apparatus. The chambers are filled with buffer and the gel is placed on the center platform, with the buffer just covering gel. Samples are loaded into preformed wells in the gel, and the power supply is attached to the electrodes for running gel.

METHODS

1. Heat agarose stock in microwave and cool to 50°C. Remove 15 ml with a 25-ml pipette (volume may vary with gel mold purchased). Excess stock solution can be reheated for later use; additional H_2O may be necessary to maintain volume.
2. Gently dispense over clean minigel plates. The full 15 ml should stay on the plate by surface tension. Edges can be taped, if desired; newer plastic gel molds have preformed well holders for sample makers.
3. Add comb with well former at proper depth, 1 mm above plate surface.[1]
4. Let gel harden (takes approximately 20 min). Remove comb.
5. Set gel in minigel apparatus with 500 ml of 1× TAE buffer containing 50 μl of ethidium bromide stock solution.[2]
6. To 8 μl of sample containing 100–500 ng of DNA in a microfuge tube, add 2 μl of gel loading solution.[3] Spin briefly to mix and collect sample.

---

[1] Homemade systems without holder for sample well former require that a well-forming comb be positioned by attaching it to the bridge with fingers 2 mm above plate before pouring gel. These plates are usually 7 × 8 cm in size.

[2] Ethidium bromide is a mutagen. Use care in handling gel and buffer. Wear gloves.

[3] If λ DNA or RE-digested DNA is being electrophoresed, heat sample at 65°C for 2 min before loading to gel to denature the cohesive ends.

7. Apply a 10-μl sample to each well.
8. Include as a standard 4 μl of *Hin*dIII-cut λ cI857 DNA (250 ng per microliter with 1 μl loading buffer) to one lane as a size marker. (Standard bands will be 23.1, 9.4, 6.6, 4.4, 2.3, 2.0, 0.56, and 0.12 kb.)[4]
9. DNA will run from the minus to the plus pole.
10. Run at 80 V for approximately 1 hr or until bromophenol blue dye front is 75% across the gel.
11. Remove gel. Rinse with H_2O to remove excess ethidium bromide. Take photo of gel on UV transilluminator to visualize bands. After photography, gel can be run further if more resolution or information is needed. Alternatively, restriction fragments may be electroeluted from the gel for further analysis (Section 9-3).

REFERENCE

Sanger, F., et al., J. Mol. Biol. *162*:729, 1982.

---

[4] See note 3.

SECTION **9-2.**

Analysis of DNA Fragments After Enzymatic Cleavage: Agarose Gel Electrophoresis

DESCRIPTION

This method describes some basic principles of agarose gel electrophoresis performed on a larger scale. The larger-scale gel allows for a longer electrophoresis and better resolution of fragments. In addition, it can resolve fragments from preparative digests of plasmids for making DNA probes. These gels are well suited for Southern blotting. Their disadvantage compared to minigels is the longer electrophoretic time required.

TIME REQUIRED

4 hr to overnight

SPECIAL EQUIPMENT

Horizontal agarose gel (submarine) electrophoresis apparatus, well-forming combs, gel mold, etc. (e.g., BRL model H4)

REAGENTS

40× TAE buffer

Ethidium bromide, 10 mg/ml

Gel loading solution

METHODS

In Advance

Prepare an RE digest of the DNA samples to be analyzed (Section 5-4). If appropriate, check progress of digestion with a minigel (Section 9-1).

Agarose Gels

1. If RE digest is satisfactory, a larger size agarose gel can be run with samples. Casting of the gel is described in detail in Section 5-5. This is a 300-ml, 20 × 20 cm gel. The agarose content used will vary with the size of the desired DNA pieces. A gel with 0.8% agarose will resolve a good range of fragment sizes (e.g., 1–25 kb). For larger fragments (e.g., 20–100 kb), use 0.5% agarose. For small fragments (e.g., 0.2–2 kb), prepare 1.5% agarose.
2. In addition to fragment separation, agarose gels can be used to help determine the size of DNA fragments. The rate of migration of DNA fragments is inversely proportional to the logarithm of their length in base pairs. Comparison of the distance migrated in an experiment with comigrated fragment size standards allows reasonably accurate size determination (± 5%) of the fragment of interest.
3. Size markers are generated from RE digests of well-characterized plasmids or bacteriophages. The sizes of the cleavage products for the standard are known and can be used to estimate the size of unknown DNA fragments by establishing a size calibration curve on semilog graph paper. The size (in kilobases) is plotted on the logarithmic ordinate, and the distance migrated (in centimeters) is plotted on the linear abscissa to make this curve. Some common size markers are as follows:[1]

   a. *Hin*dIII-cut λ DNA (λcI857Sam7) gives eight fragments: 23.1, 9.4, 6.6, 4.4, 2.3, 2.0, 0.56, and 0.12 kb.
   b. *Taq*I-cut pBR322 gives seven fragments ranging from 0.141 to 1.444 kb: 1444, 1307, 475, 368, 315, 313, and 141 bp.
   c. *Bam*HI-cut λcI857Sam7 DNA gives six fragments: 16.8, 7.2, 6.8, 6.5, 5.6, and 5.5 kb.

4. To 1–2 μg of size markers, add one-fifth of 1 volume of loading buffer, as is done with sample. Heat to 65°C for 2 min and load up to 35 μl per well on gel.
5. Run gel for 4 hr to overnight at 40 V in 1× TAE buffer with ethidium bromide in the agarose gel (Section 5-5).[2] Resolved restriction fragments can be viewed on a UV transilluminator box and photographed (Section 20-4).
6. Gel can now be used for Southern blot or fragment purification. A Southern blot and hybridization (Sections 5-6 and 7-2) can be done to identify DNA

---

[1] Because ethidium bromide intensity is proportional to the length and quantity of DNA, the amount of the insert can be estimated by comparing the sample intensity to the intensity of different sized markers.

[2] Ethidium bromide is a mutagen. Handle gel appropriately and wear gloves.

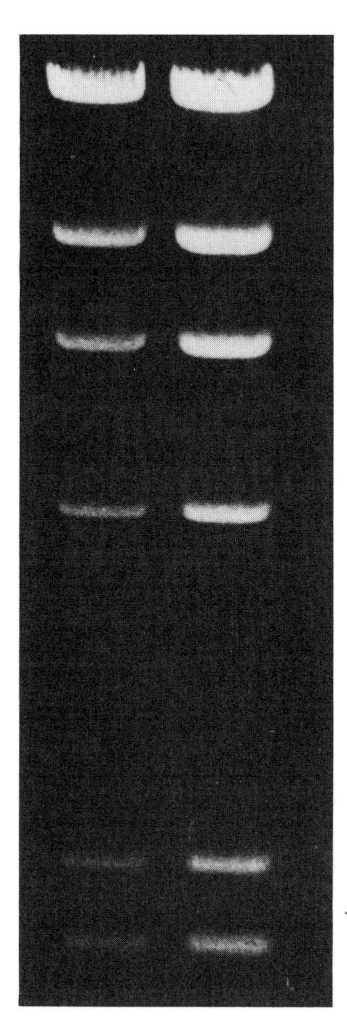

Figure 9.2
Separation of DNA fragments from *Hin*dIII digest of phage λ (cI857), at two concentrations. Photograph was made of ethidium bromide stained samples, using UV transillumination. (Photo courtesy of New England Biolabs.)

fragments that hybridize to a labeled probe. Alternatively, to purify a desired restriction fragment, cut out desired band from gel. The sample is now ready for electroelution of DNA (Section 9-3) and subsequent purification.

REFERENCES

Aaij, C., and Borst, P., Biochim. Biophys. Acta *269:*192, 1972.
Sharp, P., Sugden, B., and Sambrook, J., Biochemistry *12:*3055, 1973.

SECTION

9-3. Electroelution

DESCRIPTION

In this method, DNA fragments are recovered from agarose or polyacrylamide gels. The gel piece containing the fragment is excised and placed in a dialysis bag with buffer. Electrophoresis causes the DNA to migrate out of the gel into the dialysis bag buffer. The fragment is recovered in this buffer and purified from contaminating material using SS-phenol/chloroform extraction with or without Elutip or Nensorb chromatography. Fragments purified in this way generally work well in subsequent ligation reactions and are cleaved completely with most REs after electroelution. The method is simple, rapid, and yields a good recovery of fragment from the gel piece.

TIME REQUIRED

3–4 hr

SPECIAL EQUIPMENT

Minigel apparatus and power supply
Vacuum centrifuge (to dry samples)

REAGENTS

1 mM EDTA, pH 8.0
Na_2HCO_3, 2%
0.2× TBE buffer
SS-phenol
Ethanol
Chloroform

TE buffer
3 M sodium acetate, pH 7.4
Dialysis membrane, 13-mm diameter (Spectrapor from Spectrum)
Dialysis clips

METHODS

In Advance

Prepare dialysis tubing:

a. Cut 25-cm lengths of 13-mm-diameter dialysis tubing.
b. Add tubing to 2% Na_2HCO_3 with 1 mM EDTA. Heat until boiling, and boil for 1 min.
c. Allow to cool. Rinse tubing three times in H_2O.
d. Rinse tubing twice with ethanol.
e. Rinse tubing twice in 1 mM EDTA, pH 8.0.
f. Store at 4°C in 1 mM EDTA, pH 8.0.

Take picture of gel before cutting.

1. Cut out desired bands (visualized by UV light excitation of ethidium bromide) from the agarose gel (Section 9-2) or acrylamide gel (Section 9-4). Use an *unused* safety razor blade for each band to be cut out.
2. Place bands in individual dialysis bags (with one end of bag already clipped or tied) and add 450 μl of 0.2× TBE buffer. Squeeze out all air from bag and seal with a second clip. Cut off excess tubing. Check bag for leaks.
3. Place bags in a horizontal agarose gel or minigel apparatus, filled with 0.2× TBE buffer, just sufficient to cover the bag. Place bag perpendicular (lengthwise) to line between electrodes. Use numbered clips or string tags to keep track of DNA samples in bags if more than one bag is used.
4. Electroelute at 300 V for 1–3 hr.[1] DNA will come out of gel and remain in bag.[2] Reverse polarity and run for an additional 1–2 min to dislodge DNA from bag wall.
5. Carefully take solution out of bag with small pipette tip and place in 1.5 ml microfuge tube. Add an additional 450 μl of 0.2× TBE buffer to rinse out bag, and combine with solution. Be careful to avoid removing gel pieces.

---

[1] Electroelution can be performed at 4°C to prevent warming of sample.

[2] Because DNA is stained with ethidium bromide, you can use UV light source to visualize DNA-exiting gel. Elution time can be determined accordingly. Large DNA fragments will take longer to exit gel than short fragments. Ethidium bromide will eventually diffuse out of bag, but DNA will be retained.

6. Each sample is now approximately 900 μl. Spin in microcentrifuge for 15 sec to pellet any transferred gel pieces. Transfer supernatant in two 450-μl aliquots to two 1.5-ml microfuge tubes.[3]
7. Extract DNA with SS-phenol/chloroform and precipitate with ethanol (Section 20-1), or purify with Nensorb columns (Section 10-3).
8. Dry precipitate under vacuum in a vacuum centrifuge. This takes approximately 5 min.
9. Add 25 μl of TE buffer to each tube. Combine the two fractions from each band.

Analysis

10. To determine the approximate DNA concentration, run a few microliters of sample mixed with 1 μl of gel loading buffer on minigel. Also load known concentrations of standard-size markers (Section 9-3). Comparing the intensity of ethidium bromide staining between the eluted fragment and the standards allows an approximate determination of the concentration of the fragment. In addition to quantitation, this procedure will verify the purity of the eluted fragment.

Storage

11. The sample is very stable when stored at −20°C. It can now be used for ligation into plasmid and M13 vectors or nick translation.[4]

---

[3] If expected yield of DNA per tube is less than 5 μg, add 1 μl of 10 mg/ml yeast tRNA carrier to minimize loss.

[4] If fragments do not run well in subsequent procedures, gel-related impurities can be removed by Nensorb or Elutip methods (Section 10-3).

SECTION **9-4.**

Polyacrylamide Gel Electrophoresis of DNA Restriction Fragments

DESCRIPTION

This method is used to resolve small DNA fragments (e.g., 50–2,000 bp) and accurately determine their length by comparing them to fragments of known length. It is useful in RE site mapping with enzymes that generate small fragments. It is also useful for preparing these smaller DNA fragments by gel purification and subsequent electroelution (Section 9-3). Fragments prepared by this method work well in subsequent ligation or nick translation reactions (Section 7-1).

TIME REQUIRED

4 hr

SPECIAL EQUIPMENT[1]

Electrophoresis power supply

Gel plates and spacers for vertical gel

Vertical gel apparatus

Well formers

---

[1] These items are commerically available from a variety of suppliers or can be cut to desired sizes in a glass and plastic shop, as per individual design.

REAGENTS

10% acrylamide solution

95 g acrylamide[2] (Biorad)

5 g bisacrylamide[3] (Biorad)

H_2O to 1 liter

Dissolve by stirring while heating to 50°C. Store in the dark at 4°C.

20× TBE buffer (store at room temperature)

10% (wt/vol) ammonium persulfate (Biorad). Make fresh each week and store at 4°C.

TEMED (N,N,N′,N′-tetramethylethylenediamine) stored at 4°C

Acrylamide gel loading solution

50% glycerol

1× TBE

0.5% bromophenol blue

0.5% xylene cyanol

Ethidium bromide, 10 μg/ml

METHODS

In Advance

Prepare DNA sample to be analyzed. The sample could be RE digest of bacteriophage λ or cloned DNA in a plasmid vector (Section 5-4), or cDNA insert cut with *Eco*RI for size fractionation (Section 14-2), for example. Sample volume should be between 10–100 μl.

Prepare DNA size standards. For a typical polyacrylamide gel electrophoresis, 5 μg of pBR322 cleaved with *Eco*RI and *Hin*fI forms a good set of markers (998, 634, 517, 506, 396, 344, 298, 221, 220, 154, and 75 bp). For larger fragments, 5 μg of *Hin*dIII-cut λcI857 DNA may also be added (23.1, 9.4, 6.6, 4.4, 2.3, 2.0, 0.56, and 0.12 kb).

Assemble plates and spacers for casting the gel. Spacers may be sealed to a watertight fit by using a thin layer of petroleum jelly or vacuum grease. Alternatively, a melted 1% agarose solution can be applied, using a syringe, around spacers. Spacers are clamped in place to secure the position of the plates.

Determine volume of chamber between gel plates and spacers, if unknown. Verify that chamber is watertight.

Procedure

1. Prepare an appropriate gel volume that is more than sufficient to fill the chamber formed between the gel plates. For 100 ml of a 5% gel mix, combine:

---

2 Wear gloves and be careful not to breathe in or touch unpolymerized polyacrylamide. This chemical is a neurotoxin.

3 Bacacrylamide (Biorad) may be substituted for bisacrylamide, if desired.

10% acrylamide solution	50 ml
H_2O	44 ml
20× TBE	5 ml

Mix by gentle stirring.

2. Immediately before adding gel mix into the chamber, add 1 ml of 10% ammonium persulfate and 75 μl of TEMED. Mix gently using a 50 ml syringe without a needle and dispense mixture into the chamber using the syringe. Fill the space entirely.
3. Promptly insert slot former into the top of the chamber and clamp tightly in place. This creates wells for sample loading.
4. Allow gel to polymerize (15 min to 1 hr).
5. After polymerization, remove bottom spacer and slot former, and mount gel in electrophoresis apparatus.
6. Fill upper and lower tanks with 1× TBE buffer, submerging upper slots. Remove any trapped bubbles from the space under the gel in the lower tank.
7. Determine volume of sample to be loaded. Add one-fifth of 1 volume of acrylamide gel loading solution.[4]
8. Load the gel by layering samples with acrylamide loading solution carefully on the bottom of the wells with a glass micropipette.
9. Run the gel by applying a voltage of 2–10 V per centimeter of gel. Run the gel long enough to resolve the sample fragments of interest. As a rough guide, the bromophenol blue dye front comigrates with DNA fragments of about 50 bp, whereas the xylene cyanol dye front runs with 400-bp fragments in a 5% gel.
10. Turn off power supply and remove the gel from the apparatus.
11. Separate the two plates with the gel adhering to one of them. The resolved DNA can now be detected by autoradiography if radiolabeled (step 12) or ethidium bromide staining (step 13).
12. Autoradiography may be performed on the "wet" gel by covering it with clear plastic wrap and applying it to XAR-5 film in a cassette. Alternatively, the gel can be transferred to Whatman 3MM paper and covered with clear plastic wrap. This gel can be autoradiographed or dried on a gel dryer and then autoradiographed (Section 20-5).
13. Ethidium bromide staining can be performed by carefully layering 50–100 ml of an ethidium bromide solution, 10 μg/ml, onto the gel while it is still adhering to one of the plates. Allow ethidium bromide solution to diffuse into the gel and stain DNA for 15 min. Gently rinse excess stain from the gel

---

[4] This solution will help sample sink to the bottom of well. Dyes will allow visual monitoring of the extent of electrophoresis.

surface with H_2O. Cover the gel with clear plastic wrap and visualize DNA by UV transillumination (with plastic wrap directly against UV source, because glass absorbs UV light). Gel may now be photographed (Section 20-4).

14. Identified fragments (from step 12 or 13) can now be cut from gel with a new razor blade and recovered by electroelution (Section 9-3).

REFERENCE

Hansen, J. N., Anal. Biochem. *116:*146, 1981.

SECTION

10

Purification of DNA

SECTION **10-1.**

Spermine Purification of DNA

DESCRIPTION

This method describes a fast, reproducible precipitation of DNA from low-salt aqueous buffers (i.e., TE or TBE). It works well with low DNA concentrations (0.1 μg/ml) and can be used to separate DNA from various impurities such as free nucleotides and proteins. The DNA fragments precipitated must be longer than 100 bp for optimal recovery. The precipitation occurs because spermine binds to DNA and condenses its structure in solution. A subsequent step in the method describes the removal of spermine bound to DNA, if necessary.

TIME REQUIRED

1–2 hr

REAGENTS

Spermine, 100 mM stock (Sigma)

HS buffer

TE buffer with **0.5 M NaCl**

Extraction buffer

75% ethanol

0.3 M sodium acetate

10 mM magnesium acetate

METHODS

In Advance

DNA to be purified should be between 0.1 and 100 μg/ml in a solution with low or no salts (salt must be less than or equal to 0.1 M). Spermine precipitation is inhibited by concentrations of Mg^{2+} over 10 mM or concentrations of EDTA over 1 mM. This limitation can be overcome by using additional spermine (see below).

Spermine Cleanup

1. Add spermine stock solution to DNA sample to attain the required final concentration, depending on salt concentration in buffer:

DNA Buffer Salt	Final Spermine Conc. (mM)[1]
Low-salt buffer (e.g., TE)	0.1
0.1 M KCl	2.0
0.01 M KCl with 6 mM $MgCl_2$ or 10 mM EDTA	1.0
0.1 M KCl with 10 mM $MgCl_2$ or 15 mM EDTA	10.0

2. Incubate on ice for at least 15 min.
3. Spin in microcentrifuge at 4°C for 10 min or in refrigerated centrifuge at 12,000 × *g* for 15 min at 4°C.
4. Discard supernatant.[2]
5. Dissolve pellet in 500 μl of HS buffer for large-scale DNA preparation (above 10 μg DNA; go to step 6) or in 25 μl of extraction buffer for small-scale preparation (go to step 7).

Spermine Extraction (Two Methods)

6. *For large-scale extraction,* place dissolved pellet in dialysis bag.[3] Dialyze for 24 hr against two changes of 100× volumes of HS buffer. Dilute DNA sample (from bag) into desired storage buffer, or precipitate DNA with ethanol (Section 20-1).

---

[1] Spermine concentration to give optimally efficient DNA precipitation. We recommend using 0.1 mM spermine with DNA sample in TE buffer.

[2] Pellet can be washed with 80% ethanol, if desired at this step.

[3] See Section 9-3 for dialysis tubing preparation.

7. *For smaller preparations,* dissolve pellet with extraction buffer. Incubate for 1 hr on ice. Vortex twice during this incubation. Spin in microcentrifuge at 4°C for 10 min. Discard supernatant. Dissolve pellet in desired buffer.

REFERENCES

Gosule, L. C., and Schellman, J. A., J. Mol. Biol. *121*:311, 1978.

Hoopes, B. C., and McClure, W. R., Nucleic Acids Res. *9*:5493, 1981.

Widom, J., and Baldwin, R. L., J. Mol. Biol. *144*:431, 1980.

SECTION

10-2. Glass Powder Elution of DNA

DESCRIPTION

This is an alternative to electroelution for purifying fragments from agarose gel pieces. The gel is dissolved in NaI, and the DNA is specifically bound to glass powder. The bound DNA is washed to remove impurities and is subsequently eluted from the glass in a low-salt buffer. This method does not work with agarose gels made in a borate buffer or with acrylamide gels (gel pieces will not dissolve). This protocol can also be used to purify DNA in solution, such as M13 DNA preparations for sequencing (Section 16-7).

TIME REQUIRED

4–5 h after solutions are made

REAGENTS

Prewashed 325 mesh powdered flint glass fines[1]

TE buffer

---

[1] Glass powder fines were available commercially (IBI) and may be available from other suppliers. The fines can also be prepared in the laboratory. This is a dangerous procedure so proceed very carefully. See Reference for details.

Procedure: Suspend 250 ml of 325-mesh powdered flint glass (Eagle Ceramics) in 200 ml of H_2O in a 500-ml glass beaker. Stir for 1 hr. Let beaker stand for one additional hour for settling of large pieces. Transfer the supernatant to 50 ml tubes and centrifuge at 5000 × g for 10 min. Resuspend pellet in 150 ml of H_2O in a glass beaker. Add one-half volume of concentrated nitric acid. Heat to almost boiling in a fume hood. BE EXTREMELY CAREFUL and protect skin and clothing. Let mixture cool. Collect acid-treated glass powder by centrifugation at 5000 × g for 10 min. Wash with changes of H_2O until the pH of the mixture is 7.0 (test with pH paper, approximately four changes of H_2O). Store fines as a 50% slurry in H_2O.

Sodium iodide solution

90.8 g sodium iodide
1.5 g sodium sulfite
Adjust volume to 100 ml with H_2O
Filter through Whatman no. 1 paper or 0.45-μm filters.
Add 0.5 g sodium sulfite to saturate.
Store in dark at 4°C (foil covered).

Ethanol wash solution

50 ml ethanol
50 ml containing:

20 mM Tris, pH 7.4
1 mM EDTA
0.1 M NaCl

Store at −20°C

3 M sodium acetate, pH 7.4
Ethanol

METHODS

In Advance

Photograph gel before cutting. Cut out desired band, labeled with radioactive material or stained with ethidium bromide, from agarose gel with a new razor blade. Wear gloves throughout procedure.

Elute from Gel

1. Dissolve agarose gel slice containing desired DNA fragment in 2 ml sodium iodide solution per gram of gel. Warm solution to 37°C for 1 hr. Keep in dark.[2]
2. Dilute glass fines to a 50% slurry (vol/vol) in H_2O.
3. When all agarose is completely dissolved, shake glass slurry until all powder is in suspension. Add 3 μl of slurry per expected microgram of DNA to the dissolved gel solution. Let stand on ice for at least 90 min.[3]
4. Spin for 1 min in microcentrifuge. Discard supernatant. Wash pellet in 500 μl of sodium iodide solution at 4°C. Spin for 1 min in microcentrifuge. Discard supernatant.

---

[2] Gels run in a borate buffer, such as TBE buffer, cannot be used, because gel slices will not dissolve properly.

[3] The exact amount of glass slurry needed to bind a specific amount of DNA may have to be adjusted for each glass preparation. It is not advantageous to use an excessive amount of glass.

5. Add 500 μl of ethanol wash solution at 4°C. Spin for 1 min in microcentrifuge. Discard supernatant. Repeat for a total of two ethanol washes.

Elute from Glass Powder

6. Add 25 μl of TE buffer to glass pellet. Incubate for 30 min at 37°C.
7. Spin in microcentrifuge for 1 min. Transfer liquid to a fresh microfuge tube. Sample can now be used directly (go to step 10) or ethanol precipitated (go to step 8).

Sample Extraction

8. Add 3 μl of 3 M sodium acetate. Mix. Add 75 μl of ethanol. Freeze on dry ice for 10 min.
9. Spin in microcentrifuge for 10 min at 4°C. Discard liquid. Dissolve pellet in TE buffer.
10. DNA content can be measured by O.D. (Section 20-3) or by comparative ethidium bromide staining on gel (Section 9-3, step 10).
11. Store sample at −20°C.

REFERENCE

Vogelstein, B., and Gillespie, D., Proc. Natl. Acad. Sci., USA *76:*615, 1979.

SECTION **10-3.**

Purification of DNA: Other Methods

DESCRIPTION

This method uses two commercially developed chromatography columns for purifying DNA from impurities. Both the Nensorb (Du Pont/NEN) and Elutip (Schleicher and Schuell) products are described and work well. Often, prepared DNA fragments that do not perform well can be improved by using these purification steps. In addition, radiolabeled triphosphates can be removed from larger DNA fragments or oligonucleotides by these techniques.

TIME REQUIRED

30 min

REAGENTS

For Nensorb-20

Nensorb column package (Du Pont/NEN)

Absolute methanol, HPLC grade

Wash solution A

0.1 M Tris, pH 7.4

Add 10 mM triethylamine and 1 mM EDTA if using for samples other than synthetic oligonucleotides.

Elution solution B:

20% Ethanol in H_2O[1]

Two syringes (6 ml)

---

[1] Other solvents can be used to elute sample from column. 70% *n*-Butanol or 50% methanol is a good solvent for this purpose. Be careful that samples previously extracted with *n*-butanol do not have trace amounts of the solvent remaining.

For Elutip

Elutip column (Schleicher and Schuell)

Two syringes (6 ml)

0.45-μm cellulose acetate membrane filter, Elutip prefilters (Schleicher and Schuell #NA 010/27)

Low-salt buffer

0.2 M NaCl
20 mM Tris, pH 7.4
1 mM EDTA

High-salt buffer

1 M NaCl
20 mM Tris, pH 7.4
1 mM EDTA

TE buffer

Ethanol

METHODS

In Advance

Obtain DNA sample to be purified.

Nensorb Method

1. Assemble column as indicated, using package instructions.
2. Wash Nensorb column with 6 ml of absolute methanol, applying gentle pressure to the syringe.[2] Follow by washing with 6 ml of solution A.
3. Load sample on column in 0.5 ml of solution A.[3] Wash column with additional 1.5 ml of solution A.
4. Elute DNA with 0.5 ml of solution B. Collect the first 0.5 ml (approximately 11 drops) off the column. All of the DNA should be in these drops.[4]
5. Dry sample in a vacuum centrifuge.
6. Resuspend in buffer of choice. Discard Nensorb column.

[2] Forcing syringe may cause loss of sample. Do not allow resin to dry between steps.

[3] A maximum of 20 μg of DNA is recommended for the Nensorb-20 columns.

[4] The Nensorb columns allow a quick purification of DNA without leaving any salt residue. Dispose of column after use.

Elutip Method

DNA to be purified must be in a low-salt buffer (e.g., less than 0.2 M NaCl) before starting. Samples can range from 10 ng to 100 μg and must be less than 5 kb in size.

1. Prepare column by loading 3 ml of high-salt buffer in a 6-ml syringe and attaching column to end of syringe. Run buffer through column using only gentle force. Do not draw back on plunger or packing of resin will be disrupted.
2. Take up 5 ml of low-salt buffer in a second syringe. Attach to column and run buffer through, as described above.
3. Load DNA sample into low-salt syringe. Attach 0.45-μm cellulose acetate filter to syringe and attach column to other side of filter. Slowly drip through the filter and column.[1]
4. Disconnect and load syringe with 3 ml of low-salt buffer. Reattach filter and column, and wash filter with this solution.
5. Load 400 μl of high-salt buffer in a syringe. Attach directly to column (i.e., no filter) and slowly force through column, collecting eluted DNA in a 1.5-ml microfuge tube.
6. Precipitate DNA by adding 1 ml of ethanol (with yeast tRNA as carrier, if needed with low DNA concentrations). Freeze on dry ice for 10 min. Spin in microcentrifuge for 10 min at 4°C.
7. Resuspend DNA pellet in buffer of choice.[2]

REFERENCES

See package material accompanying Nensorb or Elutip columns.

---

[1] Forcing syringe may cause loss of sample.

[2] H_2O, TE buffer, TBE buffer, etc. can be used.

SECTION

11

Preparation and Analysis of RNA from Eukaryotic Cells

SECTION

11-1. Guanidine Isothiocyanate Preparation of Total RNA

DESCRIPTION

This method is a versatile and efficient way to extract intact RNA from most tissues and cultured cells, even if the endogenous level of RNase is high. The cells are lysed in guanidine isothiocyanate using a tissue homogenizer (Polytron). The lysate is layered onto a CsCl gradient and spun in an ultracentrifuge. Proteins remain in the aqueous guanidine portion, DNA bands in the CsCl, and RNA pellets in the bottom of the tube (a jelly-like pellet). The RNA is recovered by redissolving the pellet. The recovery of RNA is usually excellent if the capacity of the gradient is not exceeded. This method can be used to isolate RNA from tissue (follow steps 1 to 13) or to isolate both RNA and DNA from cells (start with step 14).

TIME REQUIRED

Day 1—prepare tissue and begin gradient centrifugation, 1 hr

Day 2—purify RNA, 3 hr

SPECIAL EQUIPMENT

Ultracentrifuge and swinging bucket rotors (Beckman SW41 and SW55 or Du Pont/Sorvall TH641 and TH655) with appropriate tubes

Polytron or equivalent

REAGENTS

0.3 M sodium acetate, pH 6 (sterilize by autoclaving)

3 M sodium acetate, pH 6

tRNA, 10 mg/ml

GIT buffer: To make 200 ml:

94.53 g guanidine isothiocyanate (GIT) (4 M final)
1.67 ml 3 M sodium acetate, pH 6
Bring volume to 200 ml
Sterile filter (0.8-μm filter; Nalgene #380-0080)
Add 1.67 ml β-mercaptoethanol (add in fume hood)

CsCl buffer: To make 100 ml:

95.97 g CsCl (5.7 M final)
0.83 ml *3 M sodium acetate, pH 6*
Bring volume to 100 ml
Sterile filter (0.8-μm filter)

Absolute and 80% ethanol
Proteinase K (BM)
PK buffer

150 mM NaCl
10 mM Tris, pH 7.4
10 mM EDTA
0.4% SDS

SS-phenol
Chloroform

METHODS

Cultured Cells

1. Place approximately 5×10^7 cells in 10-ml snap-cap polypropylene tubes.[1,2] Centrifuge cells to remove medium. Resuspend pellet in 4 ml of GIT buffer per gram of cells. Go to step 4.

Mouse Organs

2. Sacrifice mouse. Rinse blood off with H_2O, if necessary. Wipe stomach with ethanol and cut open abdomen. Remove organs of interest. Those listed below are good examples of tissue that can be conveniently dissected and

---

[1] Mouse tissues or a pellet of approximately 10^8 cells can be used. Other species/organs will also work, but keep tissue size uniform, as indicated. Keep tissue under 1 g per 11-ml tube or 0.5 g per 5-ml tube, or RNA yield will drop due to overloading of gradient.

[2] Use only sterile disposable plasticware, where possible, and autoclaved solutions to avoid contamination. Autoclave microfuge tubes and other glassware. Wear gloves throughout procedure.

used. Briefly blot each tissue to remove blood. Each can be placed directly into tube for step 3 or frozen in liquid N_2 until ready.[3] Add organs to individual 10-ml polypropylene tubes with cap.[4]

Large Organs (use 0.5-1.0 g of each)	Small Organs (use 0.25-0.5 g of each)
a. Spleen	g. Pancreas
b. Kidney	h. Testes
c. Abdominal smooth muscle	i. Skeletal muscle
d. Liver	j. Thymus
e. Salivary gland	k. Heart
f. Brain	l. Lung

3. Add ice cold GIT buffer: 4 ml to tubes a to f; 2 ml to tubes g to l.
4. Homogenize each tube in Polytron (or equivalent) on ice at medium speed for 15 sec (be consistent in this step). Between samples, wash Polytron carefully with large volumes of H_2O (e.g., twice in 1-liter flasks), followed by a GIT buffer rinse (in a 15-ml capped tube),[5] to avoid cross-contamination of samples. Samples may be stored at −70°C at this point if it is inconvenient to centrifuge them immediately.
5. In ultracentrifuge tubes, add:
   a. For samples a to f or cultured cells: In approximately 10-ml tubes, add 4 ml of CsCl buffer; then carefully layer guanidine lysed sample on top (e.g., Beckman SW41 or Sorvall TH641).
   b. For samples g to l: In approximately 5-ml tubes, add 1.7 ml of CsCl buffer; layer guanidine lysed sample on top (e.g., Beckman SW55 or Sorvall TH655)
6. Fill each tube to the top with GIT buffer to prevent collapse of tube during centrifugation. Spin overnight (21 hr) at 35,000 rpm (179,000 × g) for SW55 or 32,000 rpm (174,000 × g) for SW41 at 20°C. For reproducible yields, be consistent with centrifugation times, and be careful not to chill centrifuge below 20°C or CsCl may precipitate in the gradient.
7. Remove most of the supernatant by suction.[6] Remove final 200–400 μl with hand pipette. Do not touch pipette tip to bottom of tube. Swab around tube sides without disrupting pellet to remove any remaining buffer. The clear, gelatin-like pellet at the bottom is the RNA sample.[7]

---

[3] See note 1.

[4] See note 2.

[5] It is best to proceed to step 5, but samples can be frozen at this point at −70°C.

[6] See note 2.

[7] RNA pellet may be invisible or clear, but proceed anyway. If pellet is large, white, and granular, CsCl has precipitated, probably due to centrifuge dropping below 14°C or excessive centrifuge speeds, which can be dangerous. Try to remove CsCl and recover RNA by precipitating in ethanol repeatedly.

8. Resuspend pellet by vortexing or repeated pipetting in 200 μl of 0.3 M sodium acetate, pH 6. Transfer to 1.5-ml microfuge tube. Rinse tube with additional 100 μl of 0.3 M sodium acetate and add to microfuge tube.
9. Go to step 10 unless RNA preparation appears to be contaminated at this step, (i.e., granular or discolored). Sample can be further purified: Add 300 μl each of SS-phenol and chloroform. Mix by vortexing. Spin for 1 min in microcentrifuge to separate phases. Remove upper aqueous phase to new tube.
10. Add 750 μl of ethanol. Freeze on dry ice for 10 min.
11. Spin in microcentrifuge for 10 min. Discard supernatant. Add 300 μl of 80% ethanol. Mix. Spin in microcentrifuge for 2 min at 4°C. Discard supernatant. Dry pellet in vacuum centrifuge. Resuspend each pellet in 200 μl of H_2O for tubes a to f and 100 μl of H_2O for tubes g to l.[8]
12. Determine RNA quantity by O.D. at 260 nm (Section 20-3).
13. Samples can be stored at −70°C and are stable for at least a few months.[9]

For DNA and RNA Recovery

14. Place approximately 5 × 10^7 cultured cells, suspended in media, in a polypropylene tube.[10] Centrifuge cells at 800 × *g* for 5 min. Decant the supernatant. Resuspend cells in 4–8 ml of GIT buffer by shaking or vortexing. The GIT buffer will dissolve the cells, releasing nucleic acids and increasing the viscosity of the solution.
15. Layer the dissolved cells in GIT buffer onto 4 ml of CsCl buffer in a Beckman SW41 or Sorvall TH641 tube. Fill the tube with GIT buffer.
16. Spin overnight at 32,000 rpm (174,000 × *g*) at 20°C.
17. Carefully remove the GIT solution in the upper two-thirds of the tube. With a large bore pipette, remove the CsCl buffer in the lower third (4 ml) of the tube, which contains the DNA. Do not disturb the RNA pellet at the bottom of the tube. The presence of DNA in this CsCl solution will be apparent from the increased viscosity. The clear gelatin-like pellet at the bottom of the tube is the total RNA. Process this RNA pellet as described in steps 8 to 13.
18. Place the CsCl solution containing DNA into a 50-ml polypropylene tube. Add 11 ml of H_2O. Add 30 ml of ethanol to precipitate DNA. Mix by gentle inversion. DNA should precipitate immediately at room temperature and may be recovered by spooling the stringy white DNA precipitate on a pipette and transferring it to a new 50-ml polypropylene tube.

---

[8] Make sure that H_2O is double distilled, sterile (autoclaved), and highly pure for this step.

[9] For safest long-term storage, reprecipitate with ethanol, centrifuge, and store pellet under ethanol at −70°C.

[10] See note 2.

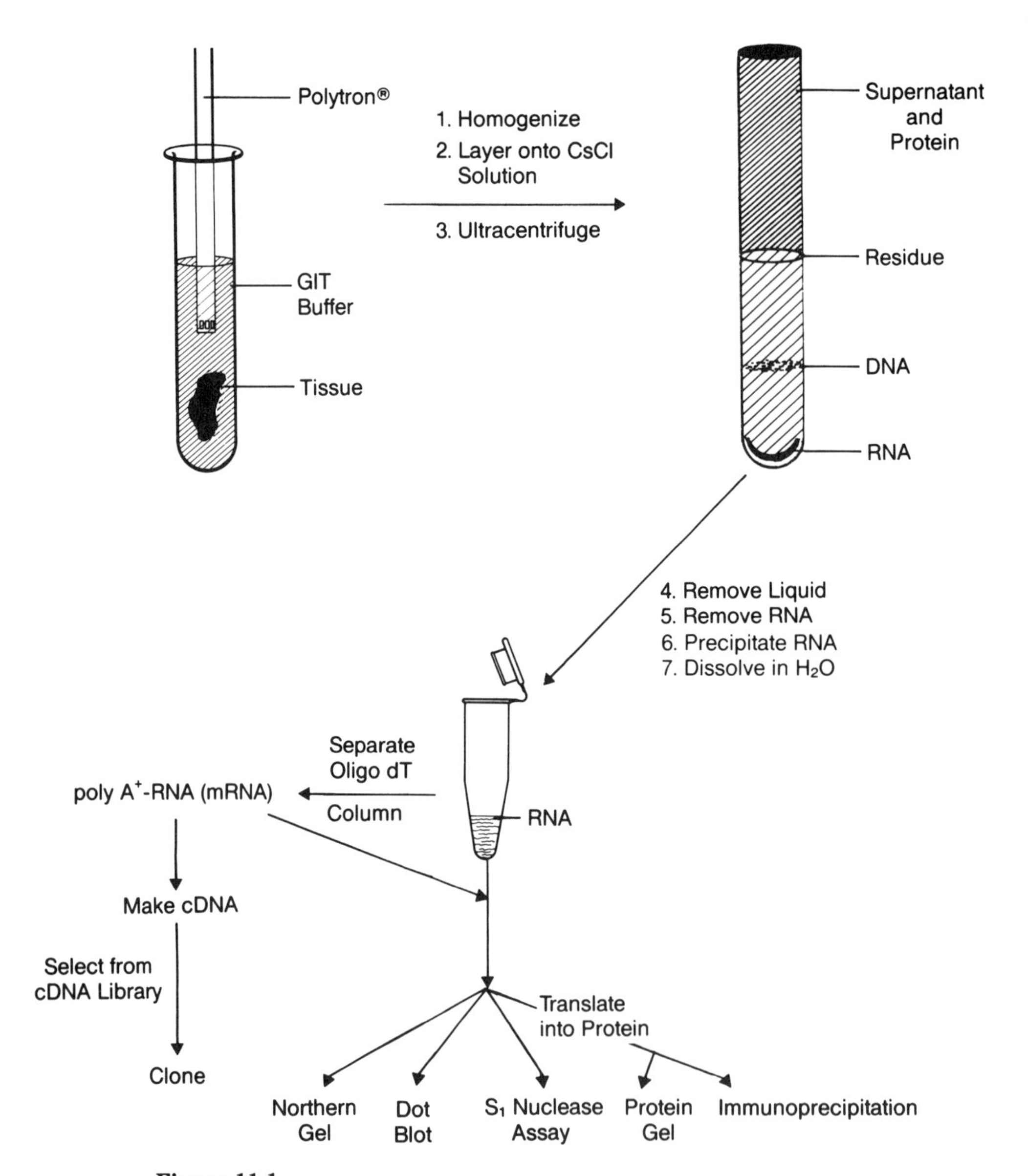

Figure 11.1
Preparation of RNA. Tissue is homogenized in guanidine isothiocyanate (GIT) and the homogenate is separated by centrifugation through a CsCl gradient. After purification, the RNA product has multiple uses, as indicated.

19. Rinse DNA pellet with 10 ml of 80% ethanol. Recover DNA precipitate by spooling or by centrifugation at 3,000 × *g* at room temperature. Decant the supernatant. Air-dry the DNA pellet, but do not allow to dry completely.

20. Resuspend the DNA pellet in 10 ml of PK buffer. If the pellet is large this step may require gentle shaking at 37°C for several hours.

21. Add 10 mg proteinase K. Incubate at 65°C for 15 min.

22. Shake gently overnight at 37°C.

23. Add 5 ml of chloroform and 5 ml of SS-phenol. Mix by gentle inversion and separate phases by centrifugation for 10 min at 12,000 x *g* in a 30-ml acid-washed Corex tube.

24. Remove upper aqueous phase to a new tube. Repeat step 23.

25. Remove upper aqueous phase to a new tube. Add 10 ml of chloroform. Mix by gentle inversion and separate phases by centrifugation at 12,000 × *g* for 10 min in an acid-washed Cortex tube.

26. Remove upper aqueous phase to a clean acid-washed 30-ml Corex tube. Add 1 ml of 3 M sodium acetate. Add 20 ml of ethanol. Incubate on ice for 20 min to form DNA precipitate.

27. Recover DNA precipitate by centrifugation at 12,000 × *g* for 10 min at 4°C. Decant supernatant.

28. Rinse pellet with 5 ml of 80% ethanol.

29. Recover DNA precipitate as described in step 27.

30. Evaporate ethanol off of pellet, but do not dry completely.

31. Resuspend DNA precipitate in TE buffer. Large amounts of DNA may require several hours to dissolve completely.

32. Measure DNA concentration by O.D. at 260 nm (Section 20-3).

33. DNA may be stored in a sterile container at 4°C and is ready for RE digestion (Section 5-4) and Southern blot analysis (Section 5-6). For long term storage of genomic DNA, a few drops of chloroform can be added to inhibit bacterial and fungal contamination.

REFERENCE

Chirgwin, J. M., Przybyla, A. E., MacDonald, R. J., and Rutter, W. J., Biochemistry *18*:5294, 1979.

SECTION

11-2. RNA Preparation: Mini Method

DESCRIPTION

This is an alternative method for making RNA from multiple tissue samples that is less time-consuming than the previous method and has the additional advantage of not requiring an ultracentrifuge. RNA obtained by this method is adequate for use in dot blot analysis for rapid screening of isolated samples. This RNA can also be used in Northern blot analysis, but the guanidine isothiocyanate method (Section 11-1) generates better-quality RNA.

SPECIAL EQUIPMENT

Polytron tissue homogenizer or equivalent

TIME REQUIRED

2 hr

REAGENTS

Homogenization buffer

100 μl of *3 M sodium acetate, pH 6*
20 μl of *0.5 M EDTA*
20 μl of vanadyl ribonucleoside complex (BRL)
100 μl of heparin solution (10 mg/ml)
580 μl of H_2O
Total volume is 820 μl.

10% SDS
SS-phenol

Chloroform
Isoamyl alcohol
Ethanol
TE buffer

METHODS

In Advance

Prepare desired tissue for RNA isolation.[1] Wear gloves and use autoclaved, acid-washed glassware for RNA preparation. Use autoclaved microfuge tubes and micropipette tips for purified RNA preparation. Autoclave all buffers (e.g., TE) used to dissolve RNA after purification (step 13).

Procedure

1. Add 100 mg of fresh tissue to 820 μl of homogenization buffer in a sterile tube. Homogenize with a Polytron at medium speed for 15 sec.[2]
2. Immediately add 100 μl of 10% SDS. Mix by vortexing.
3. Add 1 ml of SS-phenol. Mix by vortexing.
4. Add 1 ml of chloroform:isoamyl alcohol (24:1).[3] Mix by vortexing.
5. Incubate at 55°C for 5 min.
6. Incubate in ice water bath for 5 min.
7. Spin sample for 2 min at 5,000 $\times$ g to separate phases.
8. Remove upper aqueous phase to new tube. Add 1 ml of SS-phenol and 1 ml chloroform:isoamyl alcohol (24:1). Mix by shaking.
9. Spin for 2 min at 5,000 $\times$ g to separate phases.
10. Transfer aqueous phase in equal volumes to two autoclaved microfuge tubes. Add 2.5 volumes of absolute ethanol to aqueous phase. Mix. Freeze for 10 min on dry ice.
11. Microcentrifuge for 10 min at 4°C.
12. Pour off supernatant and discard.
13. Resuspend each pellet in 25 μl of autoclaved H_2O and pool samples in a new microfuge tube.

---

[1] Any tissue can be used; 100 mg of tissue or 10^7 cells are used for each tube in this procedure. Blot off blood and remove connective tissue before processing.

[2] Clean off tissue homogenizer with large volumes of water before and after each use to avoid cross-contamination. It is a good idea to wear gloves when preparing samples to avoid contamination.

[3] Use isoamyl alcohol in fume hood. Isoamyl alcohol is used to reduce the interface between phases.

14. Check RNA concentration by O.D. measurement (Section 20-3).
15. Store at −70°C. RNA is stable for 3–4 weeks.
16. Sample can now be used in dot blot procedure (Section 11-5).

REFERENCES

Lizardi, P. M., Meth. Enzymol. *96*:24, 1983.

Scherrer, K., and Darnell, J., Biochem. Biophys. Res. Commun. *7*:486, 1962.

SECTION

11-3. Selection of Poly(A^+) RNA on Oligo(dT) Cellulose

DESCRIPTION

This method allows polyadenylated RNA species (most eukaryotic mRNAs) to be purified from nonpolyadenylated RNA (rRNA and tRNAs). Because mRNA represents only a small fraction of the total RNA in a eukaryotic cell (several percent), this step represents a considerable purification. The method relies on base pairing between the poly(A^+) residues at the 3′ end of mRNAs and the oligo(dT) residues coupled to the cellulose column matrix. Nonpoly(A^+) species are not bound and are readily washed off the column. The bound poly(A^+) RNA is eluted by lowering the amount of salt in the column buffer. Poly(A^+) RNA gives better signals than total RNA in RNA gel blotting and S_1 nuclease protection experiments (Section 17-1). It is also the starting material for cDNA library construction (Section 14-2).

TIME REQUIRED

2 hr

SPECIAL EQUIPMENT

Spectrophotometer

REAGENTS

Oligo(dT) cellulose matrix (e.g., Collaborative Research Type 3 or P-L Type 7)
0.1 M NaOH with **5 mM EDTA**
Ethanol, 95 and 70%
3 M sodium acetate, pH 6, autoclaved
Ethidium bromide, 10 mg/ml

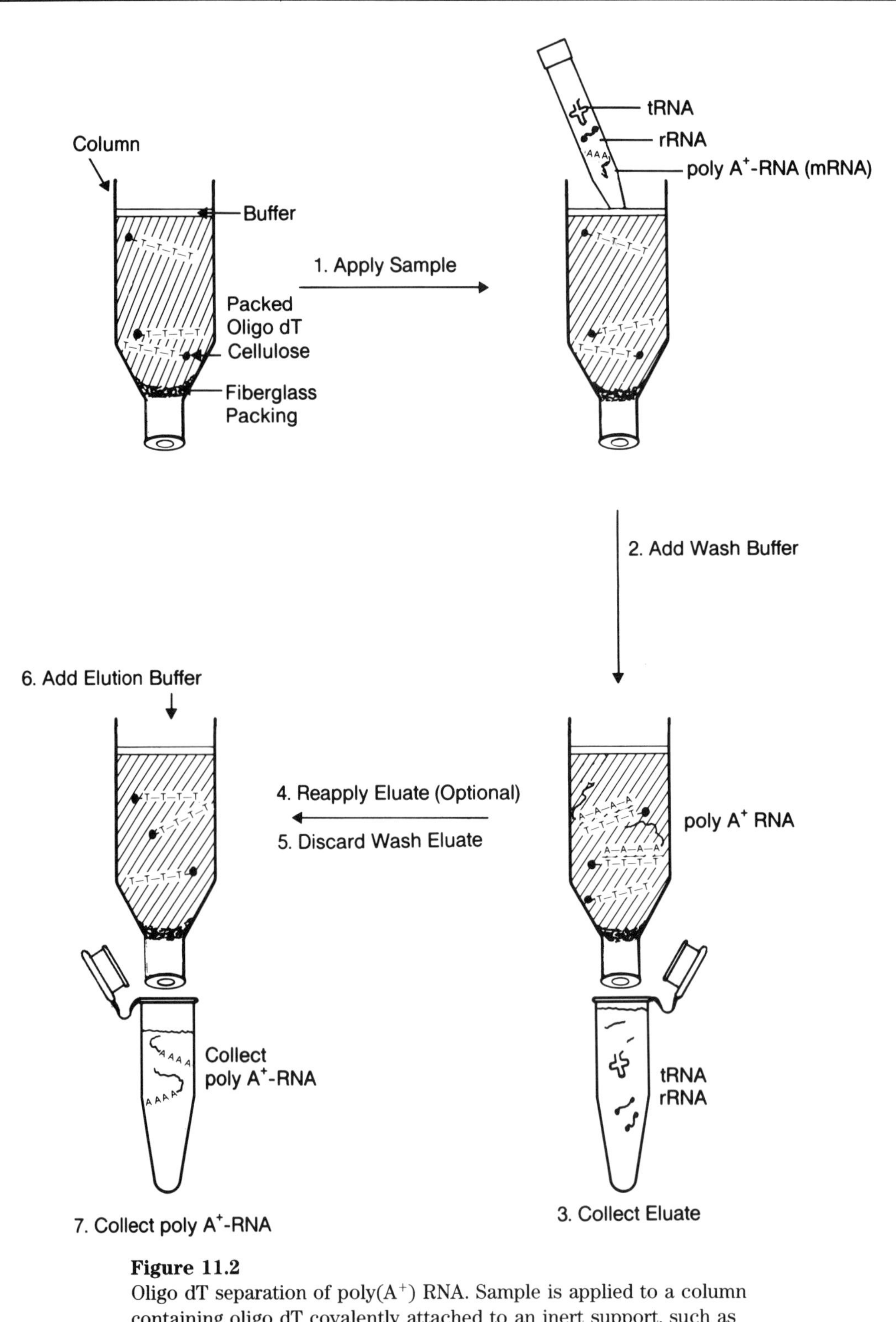

Figure 11.2
Oligo dT separation of poly(A^+) RNA. Sample is applied to a column containing oligo dT covalently attached to an inert support, such as cellulose. Only the RNA with a poly-A tail will hybridize to the oligo dT in the column, allowing the rRNA and tRNA to wash through. The poly-A^+ RNA, presumably mRNA, is subsequently eluted.

Loading buffer A

40 mM Tris, pH 7.4
1 M NaCl
1 mM EDTA
Mix and autoclave[1]
Add SDS to 0.1%

Loading buffer B

20 mM Tris, pH 7.4
0.1 M NaCl
1 mM EDTA
Mix and autoclave[2]
Add SDS to 0.1%

Elution buffer

10 mM Tris, pH 7.4
1 mM EDTA
Mix and autoclave[3]
Add SDS to 0.05%

METHODS

In Advance

Prepare column by pouring approximately 0.5 ml oligo(dT) cellulose slurry, suspended in loading buffer B, into a sterile disposable plastic column or Pasteur pipette with glass wool packing (autoclaved). The final packed column volume is typically 0.25–0.5 ml. Wash with 3 ml of 0.1 M NaOH with 5 mM EDTA. Wash column with H_2O until pH of effluent is less than 8 (by pH paper). Wash column with 5 ml of loading buffer A to equilibrate the column.

Make EB/agarose plates: Add 0.4 g of agarose to 40 ml of TE. Boil for 1 min. Cool to approximately 60°C. Add 4 μl of ethidium bromide stock solution (10 mg/ml). Pour into Petri dish. Allow to harden at room temperature.

Prepare Sample

1. Dissolve RNA sample (1–10 mg) in 500 μl of H_2O. Heat to 65°C for 5 min. Add 500 μl of loading buffer A prewarmed to 65°C. Mix. Cool to room temperature for 2 min.

---

1 Autoclave all solutions, glass column, H_2O, and so on (except SDS) to inactivate RNases. Wear gloves throughout procedure because contaminants from skin, especially RNases, may cause problems.

2 See note 1.

3 See note 1.

2. Apply this 1-ml sample to column. Collect eluate. Heat eluate to 65°C for 5 min. Cool to room temperature for 2 min.
3. Reapply eluate to column. Follow by washing column with 5 ml of loading buffer B.[4] This eluate contains nonpoly(A^+) RNA and can be discarded.
4. Add 1.5 ml of elution buffer. Collect 10-drop fractions.
5. RNA-containing fractions can be identified in ethidium bromide agarose plate assay: Spot 2 μl of each column sample onto ethidium bromide/agarose plates made earlier. Wait 20 min. Illuminate with short-wavelength UV light. All positive samples will show red fluorescence as compared to controls containing buffer alone. Serial dilutions of an RNA standard can also be run and visualized for an approximate determination of RNA concentration in sample fractions.

Purify and Concentrate

6. Pool all positive poly(A^+) RNA fractions. Add 1/10th of 1 volume of 3 M sodium acetate, pH 6. Add 2.5 volumes of ethanol.
7. Freeze on dry ice for 30 min.
8. Spin in microcentrifuge for 5 min at 4°C.
9. Discard liquid. Rinse pellet in 500 μl of 80% ethanol.
10. Spin in microcentrifuge for 3 min at 4°C. Remove supernatant.
11. Dry pellet under vacuum.
12. Resuspend in autoclaved H_2O at approximately 1 μg/μl.
13. A 2- to 4-μg aliquot of the sample can be run on 1.0% formaldehyde/agarose gel (Section 11-4) to check quality of preparation.[5]
14. Store poly(A^+) RNA at −70°C until used.

REFERENCE

Aviv, H., and Leder, P., Proc. Natl. Acad. Sci., USA *69:*1408, 1972.

---

[4] Non-poly(A^+) RNA can be recovered from the initial loading buffer eluates, if desired.

[5] One pass through column typically removes most rRNA and tRNA contaminants; two passes usually yield more than 95% poly(A^+) mRNA. The expected yield of poly(A^+) RNA should be several percent of the total RNA.

SECTION

11-4. Formaldehyde Gel for Electrophoretic Separation of RNA and Northern Blot

DESCRIPTION

The formaldehyde/agarose gel is a simple denaturing electrophoresis system that allows good size separation and resolution of single-stranded RNA. The RNA can subsequently be transferred to NC from the gel and hybridized with a radiolabeled probe of interest.

TIME REQUIRED

Day 1—5 hr to run gel and set up blot

Day 2—3 hr to bake NC filters after blotting and initiate hybridization

Day 3—2 hr to wash and set up autoradiography

SPECIAL EQUIPMENT

Horizontal gel electrophoresis apparatus

Fume hood

REAGENTS

1% SDS

Agarose (Bio-Rad or Seakem)

10× MOPS buffer (Final pH should be between 5.5 and 7.0)

- 0.2 M MOPS (3-[N-morpholino]propanesulfonic acid)
- **0.05 M sodium acetate**
- **0.01 M EDTA**

37% formaldehyde solution (Mallinckrodt)

Ethidium bromide, 10 mg/ml

Loading buffer (Make 1 ml fresh every 1–2 weeks; discard when color changes or make 10 ml and store 0.5 ml aliquots at −20°C)

0.72 ml formamide

0.16 ml 10× MOPS buffer

0.26 ml formaldehyde (37%)

0.18 ml H_2O

0.1 ml 80% glycerol

0.08 ml bromophenol blue (saturated solution)

10× SSC buffer

METHODS

In Advance

Clean gel former and all chambers, combs, and so on thoroughly with H_2O and 1% SDS or dishwashing detergent. Rinse with H_2O. It is extremely important that all components be RNase free. A separate set of equipment should be designated for this method if RNases are also used in the laboratory. Wear gloves throughout this procedure.

Prepare Gel

1. Add together:

   3 g agarose (for 1% gel)[1]

   30 ml 10× MOPS buffer

   255 ml H_2O

2. Microwave or place in boiling water bath for 5 min to dissolve and sterilize agarose gel mix.
3. Let cool to 50°C. In fume hood, add 16.2 ml of 37% formaldehyde to solution.[2] Mix by swirling.
4. Add 20 μl of ethidium bromide. Mix by swirling.
5. Let mixture sit in fume hood for 15 min to cool and to minimize formaldehyde fumes.
6. Position sample well former (comb) and pour agarose solution on plate to form gel.[3] Commercial gel former molds are convenient to use and contain positions to hold sample well formers.

---

[1] For small mRNAs (i.e., less than 2 kb), use a higher percentage agarose gel, such as 1.5% (4.5g).

[2] The original procedure called for 2.2 M formaldehyde. This concentration may adversely affect the staining of gel. The recommended concentration of 0.66 M avoids these problems.

[3] Gels are cast as described in Section 5-5.

Running Gel

7. Take 10- to 15-μg samples of total RNA [or 1–2 μg of poly(A^+) RNA] and dry sample by vacuum centrifugation.[4] Add 20 μl of loading buffer to each dry sample. Heat to 95°C for 2 min to denature RNA.
8. Place gel slab into electrophoresis tank and use 1× MOPS as running buffer. Load gel with 20 μl of RNA sample per well. Markers can be added to additional wells.[5]
9. Run gel at 200 V for approximately 2–3 hr or until bromophenol blue dye migrates three-fourths down gel. Photograph gel with ruler on UV transilluminator before blotting to record marker position. Total eukaryotic RNA contains two abundant species—28S rRNA (approximately 5 kb) and 18S rRNA (approximately 2 kb)—that can be used as molecular weight standards.

Blotting Gel[6]

10. Rinse gel for 20 min each in two changes of 500 ml of 10× SSC to remove formaldehyde from gel.
11. Fill tray with 500 ml of 10× SSC. Wet a piece of Whatman 3MM filter, approximately 23 × 50 cm, in 10× SSC buffer.
12. Place a 20 × 30 cm glass plate over tray. Drape wetted filter over plate, with both ends hanging into buffer to act as wick. Smooth out any bubbles or lumps in the filter.[7]
13. Carefully place gel upside down (i.e., sample well openings down) over wick. This can be done by placing a second plate over gel and carefully turning over the sandwiched gel. Remove new top plate and slide gel off new bottom plate onto wick.
14. Wet NC filter in H_2O. Place on top of gel. Smooth.
15. Wet two pieces of Whatman 3MM paper in H_2O. Lay individual filters flat on top of NC filter and smooth.
16. Place six paper towels flat over 3MM paper.[8]
17. Stack four piles of folded paper towels, each 2–3 cm thick, over the six towels.

---

[4] If the required volume is under 5 μl, the loading buffer can be added directly without drying the sample for a total volume of 20 μl.

[5] Markers can be made or purchased (e.g., 28S and 18S rRNA from total eukaryotic RNA, which are 5.1 and 2.0 kb, respectively, or 16S and 23S *E. coli* rRNAs, which are 1.6 and 2.9 kb, respectively). Total RNA samples already contain 28S and 18S rRNA as internal standards.

[6] See illustration on page 64 for details of blot.

[7] Smooth to remove air bubbles and excess fluid by rolling with a 5-ml glass pipette.

[8] Opened single-fold towels work well (e.g., Scott Paper #175 brown).

18. Cover with a 20 × 20 cm glass plate. Place a small weight on top.
19. Allow transfer to NC by capillary action to proceed for at least 12 hr at room temperature.

Next Day

20. Remove paper towel stacks and paper sheets. Cut off wick ends by slicing against edge of plate. Gently flip over filters and gel.
21. Remove wick and discard. Make pinholes through gel and filters, in the middle of sample well areas, as markers for electrophoretic origin. Mark first well with two pinholes to determine filter orientation.
22. Remove gel and discard. Place NC filter between two sheets of Whatman 3MM filter. Tape at two edges. Dry for 2 hr under vacuum at 80°C.[9]
23. Filter is now ready for hybridization.

Hybridization and Autoradiography

24. Hybridize filter with radiolabeled probe, either nick translated (Section 7-1) or synthetic probe (Section 6-2), as described for Southern hybridization (Section 6-3 or 7-2). There is essentially no difference in procedure for hybridizing Northern or Southern blots (Section 11-6).
25. After washing, expose filter to X-ray film overnight (Section 20-5). Develop film. Adjust time for any additional exposures, if needed.

REFERENCES

Lehrach, H., Diamond, D., Wozney, J., and Boedtker, H., Biochemistry *16*:4743, 1977.
Thomas, P., Proc. Natl. Acad. Sci., USA *77*:5201, 1980.

---

[9] Make sure filter is blotted dry before placing in vacuum oven.

SECTION

11-5. "Dot Blot" Hybridization of Labeled Probe to DNA or RNA Samples

DESCRIPTION

This method provides a rapid and less-time consuming determination of the presence of specific RNA or DNA in samples than Northern or Southern blot analyses. A direct transfer of RNA or DNA to "spots" or "dots" on NC is made. The spotted filter is then hybridized with a labeled probe. This method can also be used on partially degraded RNA samples with good semiquantitative results.

TIME REQUIRED

3 hr to prepare filter

SPECIAL EQUIPMENT

Dot blot apparatus (e.g., Bio-Rad or Schleicher and Schuell)
Vacuum source

REAGENTS

20× SSC buffer
Dye: 0.25% bromophenol blue in 25% Ficoll
TE buffer

METHODS

In Advance

Prepare DNA or RNA samples (1–2 μg) as described in Sections 5-1, 5-2, 5-3, 11-1, or 11-2. Sample should be in H_2O or TE buffer. Prepare ^{32}P-labeled probe by nick translation (Section 7-1) or end labeling (Section 6-2).

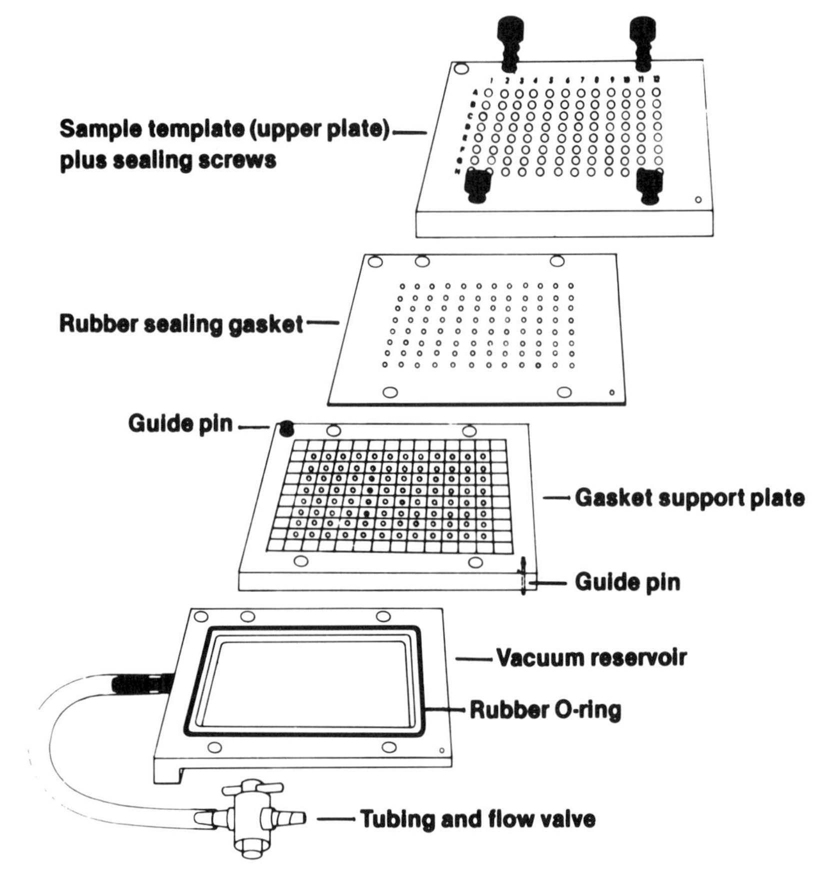

Figure 11.3
Diagram of a typical "dot blot" apparatus. The filter is placed above the sealing gasket and samples are applied to individual wells. (Illustration courtesy of Bio-Rad.)

The Blot with Apparatus (see Illustration)

1. Mix 10 ml of 20× SSC with 10 ml of H_2O.
2. Wet one piece each of NC and Whatman 3MM paper (each cut to size of apparatus) in mixture from step 1.[1]
3. Place Whatman 3MM filter in apparatus first. Place NC on top of paper.[2]
4. Put lid on apparatus and latch in place. Hook up to vacuum source.
5. Rinse wells with 100 μl of 20× SSC.
6. Place 2 μl of dye in several outer wells as position markers.

---

[1] Edges of filters may have to be cut to fit around pegs in apparatus.

[2] Be sure to clean apparatus thoroughly before and after use.

7. *For RNA dot blot:* Place 1–2 μg of RNA in 50 μl of 20× SSC. Add to wells. *For DNA dot blot:* Place 1–2 μg of DNA in 10 μl of TE buffer. Heat sample to 95°C for 5 min. Add 40 μl of 20× SSC to samples. Add to wells.[3]
8. Rinse each well with 100 μl of 20× SSC under vacuum.
9. Turn off vacuum. Take off top of apparatus and remove filter.
10. Bake NC filter between two fresh Whatman 3MM filters in vacuum oven at 80°C for 2 hr.
11. Hybridize filter with radiolabeled probe, as described in Section 6-3 or 7-2.

Without Apparatus

12. Wet NC filter with 20 ml of a 1:1 mixture of 20× SSC and H_2O.
13. Blot filter with dry Whatman 3MM paper filters.
14. In a defined pattern of rows and columns, spot small, uniform-volume DNA or RNA samples in 20× SSC buffer, as prepared in step 7, on filter. Mark additional spots with dye for orientation.
15. Perform steps 10 and 11 above.

---

[3] For semiquantitative analysis, serial 1:1 dilutions of samples added across rows can be used to yield a concentration curve.

SECTION

11-6. Probing RNA Gels: General Notes

DESCRIPTION

After a Northern blot is completed, RNA species can be identified by hybridization procedures similar to those used for Southern blot probing. Two types of probes are typically used: cloned nick-translated restriction fragments and synthetic oligonucleotides. The conditions for hybridization are identical to those described in Sections 6-3 and 7-2.

Ribosomal RNA, found in abundance (over 80% of the total RNA by mass) in cells, sometimes binds the probe in a nonspecific manner. Thus, when Northern blots are probed, extreme care must be taken to ensure that putative mRNA hybridizing signals (positive bands) migrating near the 18S (2-kb) or 28S (5.1-kb) rRNA regions of the gel are not signals due to nonspecific binding to rRNA. To minimize the chances of falsely identifying a band that is nonspecifically binding to rRNA or dismissing a positive band that runs near the rRNA region, the following steps can be taken:

1. Photograph ethidium bromide RNA gel along with a ruler carefully placed in the photographic field. This will allow fine resolution of the migration distance of bands of interest. Also, clearly mark the top of each lane with a pinhole for later calculation of the migration distance.
2. For initial runs of uncharacterized new probes, run a negative control lane containing rRNA or total RNA from cells that do not express the mRNA of interest. The affinity of the probe for rRNA can be evaluated.
3. The rRNA can be removed from mRNA in the RNA preparation using an oligo(dT) column (Section 11-3). Electrophorese total and dT-selected RNA in parallel on the RNA gel. Spurious cross-reactivity between rRNA and the probe should diminish in the dT-selected sample. Hybridization to a eukaryotic mRNA will be enriched in the poly(A^+) sample.
4. Because the formation of the double-stranded polymer is a kinetic reaction, the dissociation behavior of the bound oligonucleotide species can be used to determine whether a band is real. By washing a series of RNA-containing

filters in 0.5× SSC at different temperatures ranging from 25°C to a maximum of 80°C and autoradiographing each, the radiolabeled probe is removed.[1] The shape of the "melting curve" (formed by plotting the temperature against the signal intensity of the band) should be sigmoidal with the observed melting temperature if the band is real. Moreover, given this sigmoidal curve, the percent mismatch of an unknown RNA can be estimated (see Section 6-3).

Identification of sequences complementary to the probe is influenced by the percentage of homology, percentage of G-C pairs, salt concentration, and percentage of formamide present in the buffer. With synthetic probes, the length of the probe will also influence the hybridization temperature and wash conditions. To use nick-translated probes, follow the procedure described in Section 7-2 and adjust the wash temperature accordingly. For synthetic probes, follow the DNA probing procedure described in Section 6-3, adjusting the hybridization temperature and wash conditions as needed.

REFERENCES

Suggs, S. V., Wallace, R. B., Hirose, T., Kawashima, E., and Itakura, K., Proc. Natl. Acad. Sci., USA *78:*6613, 1982.

---

[1] NC is unstable and/or unsafe at over 80°C.

SECTION

11-7. Preparation of RNA Probes from DNA Cloned into Plasmids

DESCRIPTION

Recently, plasmids have been constructed that contain highly specific promoter sequences for transcription in vitro by DNA-dependent RNA polymerases. Examples include SP6, T3, and T7 promoters. The promotor is adjacent to a polylinker cloning site (similar to pUC or M13). If a fragment cloned 3′ to this promoter is in the polylinker, the RNA polymerase will transcribe the cloned sequence in an in vitro reaction using the purified recombinant plasmid as a template. When labeled ribo-NTPs are included in the transcription reaction, RNA probes can be synthesized. These RNA probes, which can be made in large quantity, can be used in Northern and Southern hybridizations. Moreover, RNA-RNA hydrids are more stable than RNA-DNA hybrids. The RNA can be synthesized to very high specific activity and makes a very sensitive probe for blotting experiments. It has the added advantage of being translatable with in vitro translation systems. An additional advantage of the RNA probes is their single-stranded nature; they cannot reassociate, like double-stranded, nick-translated DNA probes. The disadvantage of RNA probes is their susceptibility to degradation by ubiquitous contaminating RNases.

REAGENTS

Vectors containing DNA-dependent RNA-polymerase promoters and related materials for the in vitro transcription reactions are available commercially (e.g., SP6 PAK from Du Pont/NEN, Riboprobe System from Promega Biotec or Bluescribe from Vector Cloning Systems).

All H_2O should be autoclaved and treated with diethylpyrocarbonate (DEPC–an RNase inhibitor; 50 μl/250 ml).[1]

0.1 M sodium acetate, pH 5

[1] Wear gloves during all steps and use sterile disposable plasticware rather than washed glassware.

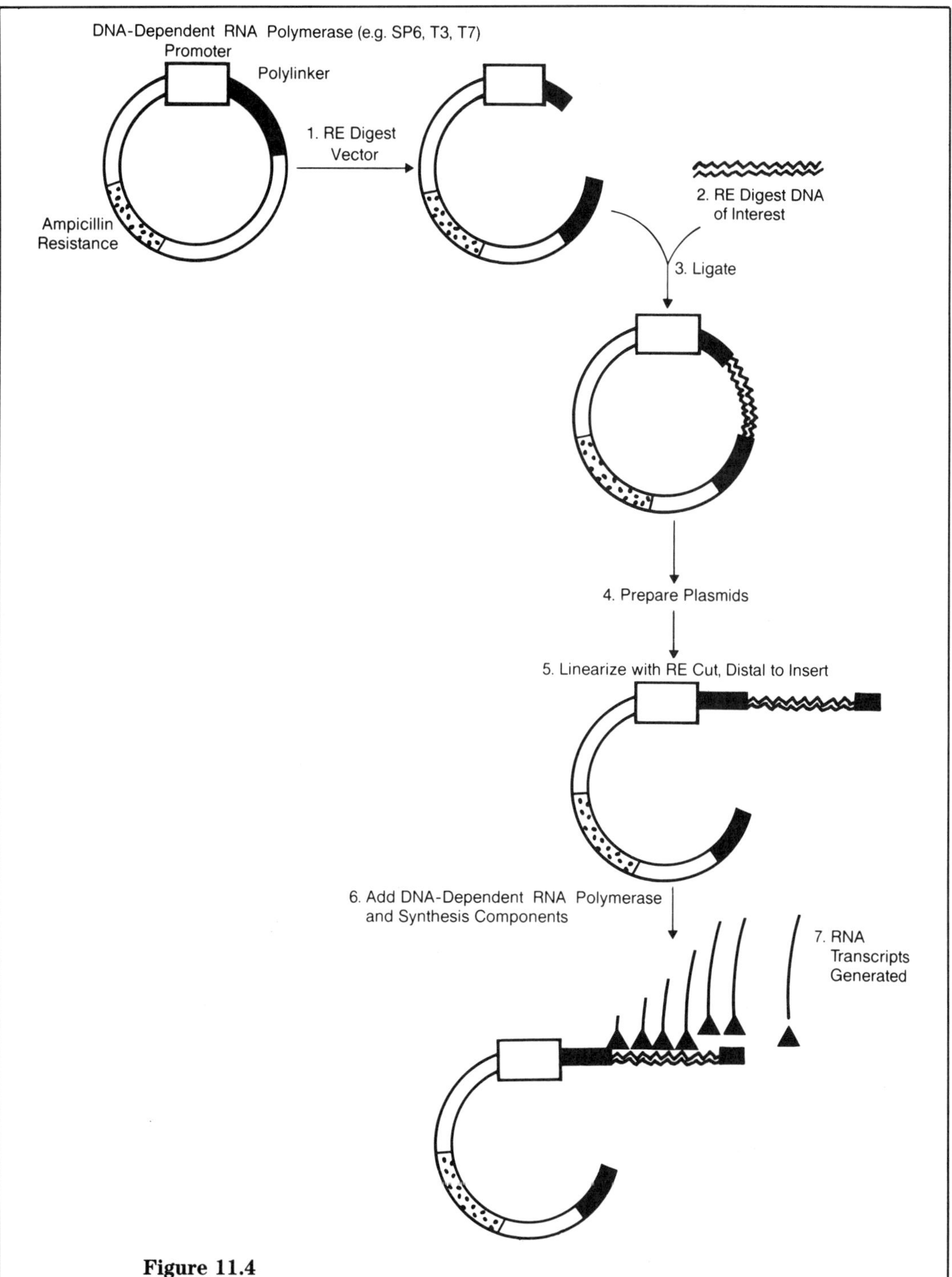

Figure 11.4
Diagram of RNA transcript formation using plasmids with DNA-dependent RNA polymerase promoters, such as SP6, T3 or T7. The DNA of interest is ligated into the vector at compatible RE-cut ends, and plasmids are prepared in the presence of ampicillin for selection. Following linearization with a RE, RNA is transcribed beginning at the promoter region using a DNA-dependent RNA polymerase. (▲=Polymerase, —— =Newly synthesized RNA strands)

Reaction buffer
200 mM Tris, pH 7.4
30 mM $MgCl_2$
20 mM spermidine

0.4 mM DTT in 50 mM sodium acetate, pH 5
2% SDS containing **2 mM EDTA**
α-^{32}P-UTP, 750 Ci/mmol (or α-^{35}S-UTP)
DNA-dependent RNA polymerase for promoter (e.g., SP6 from NEN or Promega or T3/T7 from Vector Cloning Systems)
U° is 10 mM each ATP, CTP and GTP in 50 mM Tris, pH 7. Store at −20°C.
NTP chase is 10 mM each of ATP, CTP, GTP, and UTP in 50 mM Tris; pH 7
SS-phenol
Chloroform
Ethanol

METHODS[2]

1. Insert a DNA fragment into the polylinker cloning site of the vector containing the DNA-dependent RNA polymerase promoter.[3,4] For example, the short (51-bp) universal cloning sequence has 11 specific RE sites in pSP64 or pSP65. Use the methods described for plasmid subcloning (Sections 15-1 to 15-3).
2. Prepare template by growing and amplifying plasmid in an appropriate bacterial host (e.g., LE392, HB101, or JM109) (Sections 8-1 and 8-2).
3. Linearize recombinant plasmid template with an RE that cleaves 3′ to the cloned fragment insert that is usually in the distal portion of the polylinker. Digest 2–3 μg of plasmid with an RE (Section 5-4) that will cleave distal to the promoter and the 3′ terminus of the insert. Following digestion, the plasmid is extracted with SS-phenol/chloroform and precipitated with ethanol (Section 20-1). Resuspend the plasmid at 0.2 μg/μl H_2O.

---

[2] See note 1.

[3] SP6 plasmids are pUC derived, with the SP6 promotor inserted next to the cloning sites. Currently, directional cloning is possible in pSP64 and pSP65, with the universal cloning sites running in the opposite direction in relation to the promoter. When preparing a probe to hybridize to mRNA in Northern blot or S_1 nuclease protection assays, be sure the probe transcribed from the SP6 template will be the strand that is complementary to mRNA.

[4] These plasmids have a high copy number and have the ampicillin resistance gene of pBR322 for selection.

4. Combine in a total volume of 20 μl:[5,6]

   4 μl reaction buffer
   0.5 μl DTT solution
   1 μl U°
   10 μl α-^{32}P-UTP (750 Ci/mmol) or α-^{35}S-UTP
   0.9 μg (in 4.5 μl H_2O) of the RE-cut plasmid

5. Add 14 U of the DNA-dependent RNA polymerase. Briefly centrifuge all liquid to bottom and incubate for 60 min at 37°C.
6. To stop reaction, add:

   20 μl H_2O
   25 μl 2% SDS containing 2 mM EDTA
   100 μl 0.1 M sodium acetate, pH 5

7. The solution is extracted with 100 μl each of SS-phenol and chloroform, and precipitated with 500 μl of ethanol (Section 20-1). Resuspend the probe in 1 ml of DEPC treated H_2O. Measure the amount of RNA synthesized by TCA precipitation of 2 μl of the resuspended probe as described in Section 7-1, step 8. The specific activity of the synthesized RNA should be about $1–2 \times 10^9$ cpm/μg. A good transcription reaction will produce 0.1–1.0 μg of RNA. Because the RNA labeled probe is single-stranded, it can be used directly in a hybridization reaction without denaturing. Do not denature RNA with alkali, as RNA is hydrolyzed at high pH.
8. The ethanol-precipitated pellet is redissolved in 100 μl of H_2O. A 2-μl aliquot is used to determine specific activity,[7] and an aliquot can be run on a denaturing polyacrylamide gel (Section 9-4) to evaluate products.[8]

---

[5] If nonradioactive transcripts are desired (e.g., for protein translation studies), use the NTP chase in place of the U° solution and add only 1 μCi (10 μl of a 1/100 dilution) of α-^{32}P-UTP to monitor the reaction. After the incubation (step 6), 20 μg of RNase-free DNase I is added for 10 min at 37°C, and then step 7 is followed. The RNA transcript is purified on and isolated from an agarose gel.

[6] RNAsin at a final concentration of 1 U/μl can be added to inhibit extraneous RNase activity in this step.

[7] Up to 10 μg of RNA probe can be synthesized from 1 μg of DNA template. Probes of up to 6 kb in length have been synthesized, although full-length transcripts are more easily achieved from smaller templates. Specific activity $1–2 \times 10^9$ cpm per microgram of probe should be attained.

[8] RNA probes are single-stranded. For use in Northern blot hybridizations, the RNA probes have up to a 10-fold increase in sensitivity over nick-translated DNA probes due to their high specific activity and lack of competition from a second strand. Do not denature probes with alkali as described for nick translation; dilute in hybridization buffer and add directly to NC filter in bag.

REFERENCES

Butler, E. T., and Chamberlin, M. J., J. Biol. Chem. *257:*5772, 1982.

Green, M. R., Maniatis, T., and Melton, D. A., Cell *32:*681, 1983.

Johnson, M. T., and Johnson, B. A., Biotechniques (May):156, 1984.

Kassavetis, G. A., Butler, E. T., Roulland, D., and Chamberlin, M. J., J. Biol. Chem. *257:*5779, 1982.

Melton, D. A., Krieg, P. A., Rebasliati, M. R., Maniatis, T., Zinn, K., and Green, M. R., Nucleic Acid Res. *12:*7035, 7057, 1985.

SECTION

12

Preparation of DNA from Bacteriophage Clones

SECTION

12-1. Growth and Preparation of Bacteriophage

DESCRIPTION

This method is useful as a relatively rapid initial characterization of multiple samples of cloned bacteriophage DNA and provides a bacteriophage stock to initiate a subsequent large-scale infection. A high-titer phage lysate (10 ml) is made for each clone of interest. Several micrograms of DNA can be prepared from a portion of the 10 ml lysate for further RE analysis and a determination of which bacteriophage clones should be grown on a larger scale. The remainder of the lysate is used to initiate a large-scale growth (Section 12-2) or can be stored at 4°C as a stock solution for the bacteriophage clone.

TIME REQUIRED

Day 1—1 hr plus 6–12 hr incubation time

Day 2—2–3 hr to prepare DNA

REAGENTS

LB medium

1 M $MgSO_4$

Chloroform

TM buffer

50 mM Tris, pH 7.4

10 mM $MgSO_4$

DNase I, 1 mg/ml, in TM buffer; dilute before use (Sigma #D-4263)

Polyethylene glycol-6000 (PEG-6000)

0.5 M EDTA, pH 8.0
5 M NaCl
SS-phenol
Yeast tRNA solution, 10 mg/ml, in H_2O

METHODS

In Advance

Obtain bacteriophage clone (e.g., EMBL3, EMBL4, or Charon 28) from a colleague or select bacteriophage clones of interest from a library (Section 13-5). Plate selected bacteriophage at an appropriate dilution to obtain single plaques on the day before 10 ml lysate is prepared (Section 20-7). Make fresh LE392 cells for plating in 10 mM $MgSO_4$ (Section 20-7) on the same day that 10 ml lysate is made.

Prepare 10 ml Lysate

1. Add 10 ml of LB medium supplemented with $MgSO_4$ (10 mM final) in a 50-ml capped, sterile tube.
2. With a sterile Pasteur pipette, transfer an agar plug containing a single plaque from the stock plate to the tube.[1] The plug will contain approximately 10^7 phage.
3. Add 50 μl of LE392 plating cells to tube.
4. Shake vigorously with good aeration for 6–12 hr. Culture should become cloudy and then clear with lysis.[2] Bacterial debris should be apparent with lysis.
5. If culture does not lyse, vary the number of LE392 cells in step 3 and retry.
6. After lysis, add 100 μl of chloroform to tube. Shake for 2 min at 37°C.
7. Spin tube for 10 min at 3,000 $\times$ g at room temperature to remove bacterial debris.
8. Remove upper aqueous phase to new tube. Add 100 μl of a sterile 1 M $MgSO_4$ solution to each 10 ml of lysate.
9. Titer the 10 ml lysate by serially diluting supernatant and plating (Section 20-7). Good lysates contain 5×10^9 to 10^{10} plaques per milliliter. Lysates are stored at 4°C.

---

[1] Vacuum of automatic pipettor will help bring a plug up into pipette. Dispense plug by reversing the pressure. A small volume of LB buffer can be used to rinse pipette to help release plug.

[2] A second tube with bacterial cells but no phage can be incubated along with the tube of interest. By comparing this cloudy control tube with the sample tube (lysing due to phage infection), clearing can be more easily visualized.

DNA Preparation from 10 ml Lysate Solution[3]

10. To 10 ml of lysate, add 10 ml of TM buffer and 320 μl of freshly made DNase I solution. Mix by gentle inversion. Do not vortex. Be careful not to contaminate pipettors and other equipment with DNases.

11. Incubate for 15 min at room temperature.

12. Add 2 ml of 5 M NaCl and 2.2 g of solid PEG-6000. Dissolve PEG completely in lysate.

13. Incubate for 15 min on ice.

14. Spin for 10 min at 12,000 $\times$ *g* at 4°C.

15. Pour off supernatant. Resuspend precipitated phage pellet in 300 μl of TM buffer. Transfer to 1.5-ml microfuge tube.

16. Add 300 μl of chloroform. Mix well. Spin in microcentrifuge for 5 min to separate phases. Remove upper aqueous phase to a new tube. Do not remove PEG interface between aqueous and chloroform phases. Repeat chloroform extraction one more time.

17. Add to aqueous phase:

15 μl of 0.5 M EDTA, pH 8.0
30 μl of 5 M NaCl

18. Add 350 μl of SS-phenol. Mix by vortexing. Spin in microcentrifuge for 5 min.

19. Remove upper aqueous phase to a new microfuge tube.

20. Add 350 μl of chloroform to the aqueous phase. Mix well. Spin in microcentrifuge for 5 min to separate phases. Remove upper aqueous phase to a new tube.

21. Add 875 μl of ethanol. Place tube on wet ice for 10 min. Spin in microcentrifuge for 5 min at 4°C. Decant supernatant.

22. Rinse pellet with 150 μl of 80% ethanol. Spin in microcentrifuge for 2 min at 4°C. Decant ethanol and dry pellet under vacuum.

23. Resuspend pellet in 50 μl of TE buffer.

24. DNA is now ready for RE digest analysis and electrophoresis (Sections 5-4 and 5-5) or for subcloning into plasmids (Section 15-1). The yield is generally 2–5 μg DNA per milliliter of lysate prepared if the initial titer of phage was about 10^{10}/ml.[4]

---

[3] Protocol is written per 10 ml of lysate prepared, but the preparation may be scaled-up or down linearly.

[4] This preparation generates a large amount of bacterial RNA that runs with the bromophenol blue dye front in agarose gel electrophoresis of restriction enzyme digests. The RNA background can be reduced by adding 1 μl of a 10 mg/ml DNase-free RNase A solution to the restriction enzyme digest of "miniprep" bacteriophage DNA. To make DNase-free RNase A, resuspend RNase A at 10 mg/ml in H_2O. Heat this solution to 70°C for 10 min to inactivate DNases.

SECTION

12-2. Large-Scale Preparation and Purification of DNA from Bacteriophage

DESCRIPTION

This method is one of many available for growing λ bacteriophage clones on a large scale. The small-scale 10-ml lysate (Section 12-1) is used to initiate a larger infection of bacteria. The bacteriophage eventually lyse the bacteria growing in culture, yielding between 10^9 and 10^{10} phage per milliliter of culture. The bacteriophage are then purified by PEG precipitation from the culture medium, and DNA is prepared from the bacteriophage. Typical DNA yields are 100 μg to 1 mg DNA per liter of culture.

TIME REQUIRED

5 days at several hours per day

SPECIAL EQUIPMENT

Ultracentrifuge (Beckman with SW41 and SW60 rotors)
Ultracentrifuge tubes (Beckman Ultra-clear, 14 × 89 mm: SW41; Ultra-clear, 11 × 60 mm: SW60)
Refractometer

REAGENTS

LB medium
LB medium with 10 mM $MgSO_4$ (10 ml of sterile 1 M $MgSO_4$ solution per liter of LB)
10 mM $MgSO_4$
Chloroform
5 M NaCl

PEG-6000 (solid)
TM buffer
50 mM Tris, pH 7.4
10 mM $MgSO_4$

CsCl
CsCl/TM solution, 0.75 g CsCl, added to 1 ml of TM buffer (refractive index of 1.3815)
0.5 M EDTA, pH 8.0
Pronase solution
20 mg pronase (Calbiochem) per milliliter of TE buffer

Ethanol
SS-phenol
TE buffer

METHODS

In Advance

On previous day, prepare a 10-ml lysate of bacteriophage (Section 12-1) and determine its titer by dilution and plating (Section 20-7). On day of infection, grow LE392 cells in 40 ml of LB medium to an O.D. of 1.0 at 600 nm.

Isolate DNA

1. Pellet LE392 cells in 50-ml polypropylene tubes at 3,000 × *g* for 10 min.
2. Decant supernatant and resuspend bacterial pellet in half of the original volume (i.e., 20 ml) in sterile 10 mM $MgSO_4$. This 20 ml contains about 3×10^{10} LE392 cells.
3. To bacteria, add approximately 10^9 phage, as determined by plating and titering the 10 ml of lysate (Section 20-7).[1]
4. Adsorb phage to cells by incubation for 5 min at 37°C.
5. Inoculate 1 liter of sterile LB medium containing 10 mM $MgSO_4$ in a sterile growth flask with the cells from step 4.
6. Shake vigorously for 10–12 hr at 37°C until lysis occurs. Clearing of solution may not be apparent, but presence of bacterial debris will be evident with lysis.
7. Add 20 ml of chloroform and 120 ml of 5 M NaCl.
8. Shake for 5 min at 37°C.

---

[1] This gives a phage-to-bacteria ratio of between 1:10 and 1:100, a good range for most EMBL and Charon 28 library clones. Other phage may require a different optimal ratio, which may have to be empirically determined.

9. Spin out bacterial debris in 500-ml bottles by centrifugation at 8,000 × *g* for 10 min at 4°C.

10. Pour aqueous supernatants into new flask, being careful not to dislodge pellet. The preparation can also be titered at this point to determine estimated final yield. At least 50% of the original titer is usually found in the final yield. Add 10 ml of 1 M $MgSO_4$ and 120 g of solid PEG to supernatant.

11. Dissolve PEG in lysate and precipitate phage for 60 min or longer at 4°C.

12. Collect phage by centrifugation at 8,000 × *g* for 20 min at 4°C in 500-ml bottles. Discard supernatant and resuspend PEG-phage pellet from the bottom and wall of bottle in 15 ml of TM buffer. Transfer suspended pellet to a 50-ml capped tube.

13. Add 15 ml of chloroform to tube. Mix vigorously to extract. Separate phases by spinning in a 30-ml glass tube (e.g., Corex) at 12,000 × *g* for 10 min.

14. Remove upper aqueous phase to a new glass tube. Repeat chloroform extraction as in step 13. Remove aqueous phase to a new tube. Bacteriophage may be purified by equilibrium gradient centrifugation (steps 15 to 17) or by velocity gradient centrifugation into a CsCl step gradient.[2] The step gradients are a little more complicated to set up, but one day's run in the ultracentrifuge is eliminated. Either method gives good purification of bacteriophage particles from *E. coli* DNA.

15. Add CsCl to extracted aqueous phase to a density of 1.5 g/ml by adding 0.75 g solid CsCl per milliliter of aqueous phase. Check density by confirming that the refractive index is 1.3815 ± 0.002. Adjust refractive index by adding H_2O if the index is high or CsCl if the index is low.

16. Band bacteriophage in CsCl density gradient by spinning in ultracentrifuge at 210,000 × *g* for 24–36 hr at 20°C (e.g., in two 14 × 89 mm Beckman

---

[2] The equilibrium CsCl gradient outlined in steps 15 to 17 can be replaced by a step gradient procedure to isolate bacteriophage that requires less centrifugation time. To make the gradient, three CsCl solutions of differing densities are necessary:

CsCl Solution	Density	CsCl (g)	TM Buffer (ml)
I	1.45	30	42.5
II	1.50	33.5	41
III	1.70	47.5	37.5

In SW41 tubes, add 1.5 ml of solution I to the bottom of the tube. Underlay this step with 1.5 ml of solution II, and underlay these two steps with 1 ml of solution III (add new layers by carefully pipetting to bottom through previously added layers).

For each milliliter of bacteriophage in TM solution (step 14), add 0.5 g of CsCl. Dissolve CsCl and layer the bacteriophage/CsCl solution on top of the step gradient, made as described above. Be sure to fill the tubes with TM buffer before centrifugation.

Centrifuge at 22,000 rpm for 4 hr. Bacteriophage form a bluish opalescent band at the interface between the 1.45 and 1.50 steps. Sometimes it is helpful to place the tube against a dark background and illuminate it from below to visualize the band. Proceed to step 18.

ultraclear tubes at 35,000 rpm in a Beckman SW41 rotor).[3] Be careful not to transfer any chloroform to tubes, because this will weaken ultracentrifuge tubes. Be sure tubes are full before centrifugation.

17. After equilibrium density banding, bacteriophage will form an opalescent sharp band, which should be visible in room light. If two bands are visualized, the upper (lower-density) band represents empty protein coats and the lower (higher-density) band is the phage. This lower, dense band is the one that should be removed.

18. Remove band by puncturing the side of the tube with a syringe and an 18-gauge needle. This procedure is similar to the method described in Section 8-2, step 22.

19. Centrifuge bacteriophage after removal one additional time to purify and concentrate further. This is performed by centrifugation in CsCl/TM solution in a smaller ultracentrifuge tube (e.g., SW60 tube at 42,000 rpm for 16 hr at 20°C). Be sure to fill the tubes before centrifugation. The tube must be of puncturable plastic for removal of band (e.g., Beckman Ultra-Clear, 11 × 60 mm). Identify opalescent phage band as described in step 17. Only one band should be apparent if preparation is optimal.

20. Remove bacteriophage by puncturing side of tube. Phage will keep for long periods of time when stored in CsCl at 4°C.

Purification of DNA from Phage[4]

21. For each 200 μl of CsCl banded phage, add:

50 μl of 0.5 M EDTA, pH 8
12.5 μl of pronase solution

22. Load mixture into a dialysis bag (dialysis tubing preparation is described in Section 9-3). Dialyze for 1 hr against of 10 mM Tris, pH 7.4, 1 mM EDTA, and 0.5 M NaCl at 37°C. Use at least 1,000 × the volume in the bag. Dialysis and pronase digestion will occur concurrently.

23. Change dialysis buffer and dialyze for an additional hour. Remove dialysate from bag with a micropipettor.

24. Extract dialysate from bag with an equal volume of SS-phenol in a 1.5-ml microfuge tube. Spin in microcentrifuge for 2 min.

25. Remove upper aqueous phase to a new tube.

26. Add an equal volume of chloroform. Mix. Spin in microcentrifuge for 2 min.

27. Remove upper aqueous phase to a new tube. Add 2.5 volumes of ethanol. Incubate on dry ice for 10 min.

28. Spin in microcentrifuge for 5 min at 4°C.

---

[3] It may be necessary to top off centrifuge tubes with CsCl/TM solution.

[4] There are numerous methods for making DNA from λ-derived bacteriophage.

29. Decant supernatant. Air-dry pellet.

30. Resuspend phage DNA in 500 μl of TE buffer. Allow several hours at 4°C for phage DNA to go into solution completely. Mix again and determine concentration of DNA with O.D. measurement (Section 20-3).

31. Store sample at −20°C until further use, such as RE mapping (Sections 5-4 and 5-5) or subcloning into plasmids (Section 15-1).

REFERENCES

Maniatis, T., Fritsch, E. F., and Sambrook, T., *Molecular Cloning: A Laboratory Manual*, Cold Spring Harbor, New York, 1982.

Yamamoto, K., Alberts, B., Benzinger, R., Lawhorne, L., and Treiber, G., Virology *40*:734, 1970.

SECTION

13

Cloning DNA from the Eukaryotic Genome

SECTION

13-1. Cloning DNA from the Eukaryotic Genome: Introduction

To analyze in detail the structure and function of genes and their adjacent regulatory regions, it is essential first to isolate the gene as a clone. To do this, one can generate a genomic DNA library from a cell that contains the gene of interest. This library will optimally contain an overlapping set of large DNA fragments that includes at least one copy of all sequences of the genome (i.e., over 1 million fragments of 15–20 kb average size). These fragments will be cloned and propagated in an appropriate prokaryotic vector, usually a bacteriophage. The recombinant phages thus generated constitute members of the library, each of which has its own cloned segment of DNA. The genomic library can be screened by hybridization to a radioactive probe specific for the gene of interest. This, in turn, allows selection of the desired bacteriophage clone from the remaining cloned sequences. The selected bacteriophage clones are then plaque purified and further selected by similar methods in subsequent steps. The purified clones are grown preparatively to obtain cloned DNA in amounts adequate for detailed analysis.

The high efficiency of cloning and the relative ease of screening complete libraries of eukaryotic genomes have made bacteriophage λ replacement vectors attractive for use in library generation. Lambda vectors are linear DNA phage genomes genetically engineered to allow substitution of internal, nonessential bacteriophage genes with a 15- to 20-kb segment of eukaryotic DNA.

If there are multiple DNA species in the genome that hybridize to the probe being used to select clones, another approach may be used to clone the particular sequences desired. Genomic blotting experiments are used initially to define the size of an *Eco*RI or *Bam*HI restriction fragment in the genome that contains the region to be cloned. A preparative amount (i.e., 1 mg) of genomic DNA is then digested to completion with the appropriate RE (Section 5-4) and size fractionated using preparative agarose gel electrophoresis. The size fraction containing the gene of interest is either eluted from the agarose gel or collected from the positive electrode of a preparative "bull's-eye" agarose electrophoresis gel (Hoefer Instruments sells an apparatus for this purpose). DNA is purified, ligated into an appropriate cloning vector (*Bam*HI:EMBL3; *Eco*RI:EMBL4), and packaged, and a size-fractioned library is generated. This subset of sequences

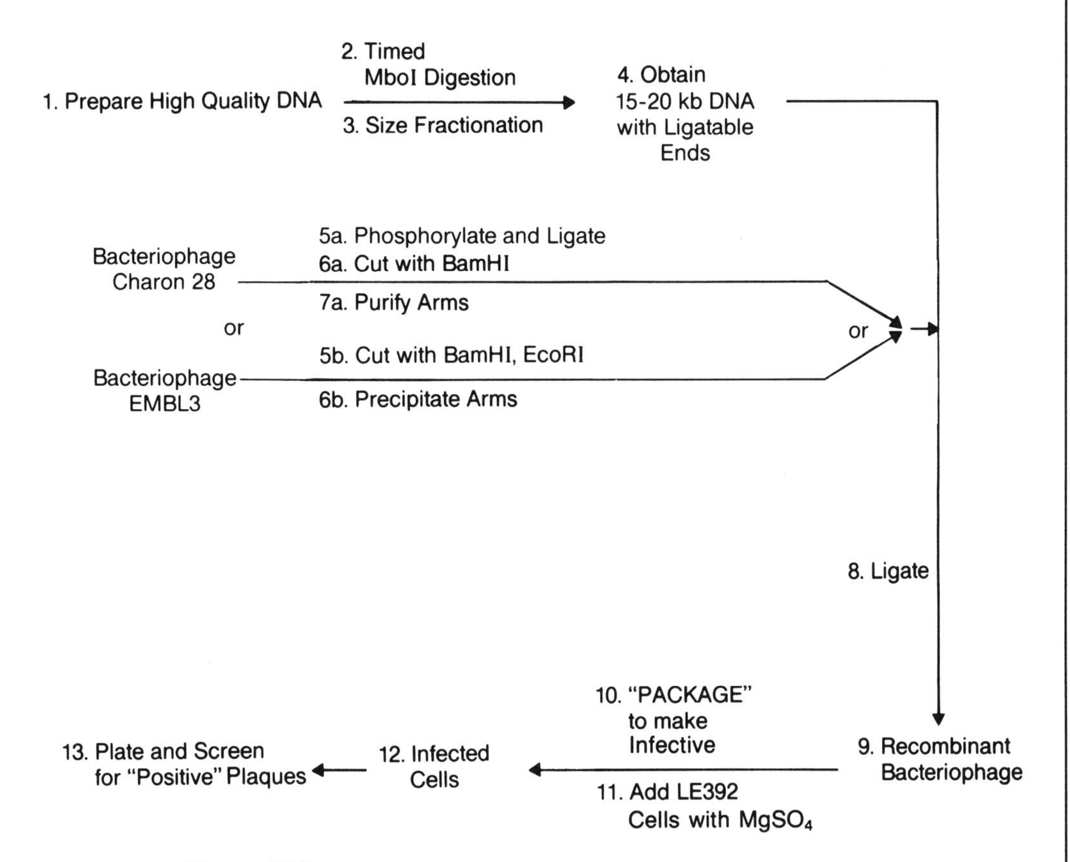

Figure 13.1
Preparation of genomic library in Charon 28 or EMBL3; flow chart.

may then be screened with the probe to clone selectively with the genomic region of interest. A summary of these procedures is presented in Figure 13.1.

A second genomic cloning vector, the cosmid, has been developed to allow the cloning of DNA segments up to 45 kb in length. Unless it is essential to clone a segment of genomic DNA larger than 20 kb, it is simplest to construct and screen λ replacement vector libraries (see references below).

The methods presented in the following sections describe procedures for generating a genomic library using a λ replacement vector. References for methods of cosmid library generation are given below. If a genomic library from an appropriate host cell already exists, skip the methods for library construction and proceed directly to methods for plating and screening (Sections 13-5 and 13-6).

REFERENCES (for cosmid cloning)

Collins, J., and Hohn, B., Proc. Natl. Acad. Sci. USA *75:*4242, 1978.

Grosveld, F., Lund, T., Murray, E., Mellor, A., Dahl, H., and Flavell, R., Nucleic Acids Res. *10:*6716, 1982.

Ish-Horowicz, D., and Burke, J., Nucleic Acids Res. *9:*2989, 1981.

Maniatis, T., Fritsch, E. F., and Sambrook, J., *Molecular Cloning: A Laboratory Manual*, Cold Spring Harbor, New York, 1982.

SECTION

13-2. Preparation of Genomic DNA: Partial *Mbo*I Digestion Method

DESCRIPTION

Large fragments from a genomic library are partially digested to 15- to 20-kb pieces with the RE *Mbo*I for cloning into the compatible RE termini of a *Bam*HI-digested bacteriophage vector. Following the digest, fragments of the desired size (15–20 kb) are separated from smaller fragments by sucrose density gradient or other methods.[1] These fragments are ligated into the appropriate bacteriophage replacement vector, packaged in vitro, and used to infect an *E. coli* host to create the genomic library.

TIME REQUIRED

Day 1—4 hr
Day 2—4 hr
Day 3—2 hr
Day 4—2 hr

---

[1] Another method for size fractionating a 15- to 20-kb partial *Mbo*I insert is to use preparative gel electrophoresis and electroelution. Alternatively, one can use a bull's-eye electrophoresis apparatus (Hoefer Instruments) that electrophoreses DNA from the outside of a circular gel into the center, where the DNA runs into a central well and is collected in a fraction collector. Samples are compared on agarose gel electrophoresis against *Hin*dIII-cut λ DNA standards. This replaces both of the sucrose density gradient steps and gives high-quality 15- to 20-kb partial *Mbo*I fragments for genomic cloning that function well in library formation. Also, this apparatus can be used to select genomic DNA fragments of other sizes for cloning. This method involves the purchase of special equipment, and we recommend that you begin with more than 1 mg of genomic DNA for the preparative fractionation.

SPECIAL EQUIPMENT

Ultracentrifuge
Sucrose gradient former
Agarose gel electrophoresis apparatus

REAGENTS

10× *Mbo*I buffer

1 M NaCl
100 mM Tris, pH 7.4
100 mM $MgCl_2$
10 mM DTT

*Mbo*I (NEBL)
Sucrose gradient, linear from 10 to 40%, with **20 mM Tris, pH 7.4, 10 mM EDTA, 1 M NaCl**
TE buffer
5 M NaCl
SS-Phenol
Chloroform
Ethanol

METHODS

In Advance

Prepare genomic DNA of best quality (undigested size over 50 kb) from tissue or cells of interest (Section 5-2 or 5-3). Test *Mbo*I.[2]

*Mbo*I Digestion

1. An analytical reaction is first run to determine the conditions of *Mbo*I digestion needed to generate DNA fragment lengths of around 20 kb.

a. Prepare four 1.5-ml microfuge tubes, each containing:

10 μg genomic DNA in 10 μl TE or H_2O	
10× *Mbo*I buffer	2.5 μl
H_2O	12 μl

b. Add 0.2, 0.5, 1, or 2 U of *Mbo*I to the four tubes.
c. Incubate at 37°C: Remove 6-μl aliquots from each tube after 5, 10, 20, and 40 min of incubation. Store each sample on dry ice until ready for next step.

---

[2] Before using *Mbo*I for insert preparation, it is advisable to verify that the enzyme will generate ligatable ends. This can be verified by cutting SV40 DNA, or plasmid grown in a *dam*$^-$ *E. coli* host, to completion with *Mbo*I and determining that T4 DNA ligase will religate the cleaved DNA to high molecular weight species.

d. Thaw and run the 16 samples on an agarose gel (Section 9-2) and compare to size markers.
e. Choose time and *Mbo*I concentration that give a mean size of approximately 20 kb.

2. Scale up the optimal digestion conditions 100 times to digest 1 mg of genomic DNA. In a 10- to 15-ml plastic tube, add:

1 mg genomic DNA	1 ml
*Mbo*I	As determined
10× *Mbo*I buffer	250 μl
H_2O	1.2 ml

Incubate for 37°C at the time determined in step 1e.

3. Extract DNA with 2.5 ml of a 1:1 mixture of SS-phenol and chloroform.
4. Remove upper aqueous phase to new tube. Add 275 μl of 5 M NaCl.
5. Add 7 ml of ethanol. Precipitate on ice for 10 min.
6. Centrifuge tube at 12,000 × *g* for 10 min at 4°C.
7. Decant ethanol. Resuspend DNA pellet in 500 μl of TE buffer. This may take a few hours at 4°C.

Size Fractionation of Insert

8. To achieve optimal size fractionation, one-sixth of DNA (approximately 80 μl) from step 7 is placed on each of six sucrose density gradients made in Beckman Ultra-clear SW41 ultracentrifuge tubes. Each linear gradient is made with a sucrose gradient former from 10 to 40% sucrose containing 20 mM Tris, pH 7.4, 10 mM EDTA, and 1 M NaCl.
9. Centrifuge at 135,000 × *g* for 16 hr at 20°C (e.g., 28,000 rpm in a SW41 rotor).

Next Day

10. Punch a small hole in the bottom of the first tube bottom and drip 12 1-ml fractions. Drip 12 fractions from each tube on top of the same fraction numbers from other tubes (i.e., the same sequential fractions from each tube are collected together).
11. Determine which fractions contain 16- to 20-kb fragments by agarose gel electrophoresis of a 50-μl aliquot as compared to *Hin*dIII-cut λ DNA markers run in parallel.
12. Pool fractions containing 16- to 20-kb fragments. Add 2.5 volumes of ethanol. Chill for several hours at −20°C.

Next Day

13. Collect DNA by centrifugation at 10,000 × *g* for 10 min at 4°C.
14. Decant supernatant. Resuspend DNA in 160 μl of TE buffer.

15. For optimal size fractionation, repeat gradient fractionation as in steps 9 to 14, except that only two gradient tubes are used with 80 μl DNA each. After the second sucrose gradient fractionation, resuspend precipitated DNA in 50 μl of TE buffer.

Next Day

16. Estimate DNA concentration and verify size fractionation of prepared insert by comparing the ethidium bromide staining intensity of a 2.5 μl aliquot of sample to that of DNA fragments of known size and concentration (e.g., 5 μg of *Hin*dIII-cut λ DNA) after agarose gel electrophoresis.
17. Adjust DNA concentration of sample to approximately 0.25 μg/μl. Store at −20°C.

SECTION **13-3.**

Preparation of Bacteriophage Vector for Genomic Cloning

DESCRIPTION

Bacteriophage vector arms (the left and right ends of bacteriophage λ) are prepared for ligation to partial *Mbo*I fragments of genomic DNA. Two vectors, Charon 28 and EMBL3, will be described.

For Charon 28, the replaceable central portion of the bacteriophage genome is removed by cleavage with *Bam*HI. Subsequent size fractionation by sucrose gradient is essential to remove the central portion, or it will compete with the cloning insert for the available vector "arms."

For EMBL3, the excised replaceable center can be inactivated in subsequent ligations by removing its cohesive *Bam*HI ends with a second RE (*Eco*RI). This is followed by removal of the excised 14-bp fragment containing the *Bam*HI cohesive ends with isopropanol precipitation. This procedure removes the central portion fragment from competition with the eukaryotic insert for ligation into vector arms, without sucrose gradient purification.

TIME REQUIRED

3 days for Charon 28

1 day for EMBL3

SPECIAL EQUIPMENT

Ultracentrifuge (for Charon 28 procedure)

Agarose gel electrophoresis apparatus

REAGENTS

Charon 28 DNA (BRL)

EMBL3 DNA (Promega or Vector Cloning Systems)

10× T4 kinase buffer
- **0.5 M Tris, pH 7.4**
- **0.1 M $MgCl_2$**
- **0.05 M DTT**
- 0.001 M spermidine

10 mM ATP, in H_2O, pH to 7.4 with NaOH. Store frozen at −20°C.
Chloroform
SS-phenol
Ethanol
T4 kinase, 17.5 U/μl (P-L)
5 M NaCl
TE buffer
Ligase buffer, 10×
T4 ligase, 100 U/μl (NEBL)
*Bam*HI (BRL or NEBL), 10 U/μl
*Bam*HI buffer, 10×
- **0.5 M NaCl**
- **0.1 M Tris, pH 7.4**
- **0.1 M $MgCl_2$**
- Store frozen at −20°C.

*Eco*RI buffer, 10×
- **0.5 M NaCl**
- **1 M Tris, pH 7.4**
- **0.1 M $MgCl_2$**
- Store frozen at −20°C.

*Eco*RI, 10 U/μl (BRL or NEBL)
3 M sodium acetate, pH 6.0 (**also 0.35 M**)
Isopropanol

METHODS

To make Charon 28 arms, the cohesive ends are phosphorylated and ligated together to form a circular genome. The central stuffer is excised by *Bam*HI digestion and is separated from the arms by sucrose gradient fractionation.

Charon 28/*Bam*HI Arm Preparation: Phosphorylation of Cohesive Ends (to Allow for Efficient Ligation)

1. Add together in a microfuge tube:

Charon 28 DNA (1 μg/μl)	500 μl
10× T4 kinase buffer	100 μl

T4 kinase	20 μl
10 mM ATP, pH 7.4	100 μl
H_2O	280 μl

2. Incubate for 1 hr at 37°C. Divide into two microfuge tubes.
3. To each tube add 1 volume (500 μl) of SS-phenol:chloroform (1:1). Mix well. Spin in microcentrifuge for 2 min to separate phases. Remove upper aqueous layer to a new tube.
4. Add 500 μl of chloroform to each aqueous phase. Mix well. Spin to separate phases.
5. Remove aqueous layer to new tube.
6. Add 50 μl of 5 M NaCl to each tube. Add 1 ml of ethanol. Freeze on ice for 10 min.
7. Spin in microcentrifuge for 5 min at 4°C.
8. Remove supernatant. Add 1 ml of 80% ethanol to rinse pellet. Centrifuge. Decant liquid.
9. Dry DNA pellet under vacuum.
10. Resuspend each DNA pellet in 250 μl of TE buffer. Pool the two samples and store at −20°C.

Ligation of Cut Ends to Each Other to Circularize Phage Genome

11. Add together:

Phosphorylated Charon 28 DNA	500 μl
10× ligase buffer	100 μl
T4 ligase	50 μl
H_2O	350 μl

12. Incubate for 4 hr at 14°C. Stop reaction by heat-inactivating enzyme at 65°C for 10 min.

Excision of Fragment

13. Add 100 μl of 5 M NaCl and 1,000 U of *Bam*HI.[1] Incubate for 2 hr at 37°C.
14. Extract and ethanol precipitate, as per steps 3 to 9.
15. Resuspend DNA in 480 μl of TE buffer.

---

[1] It is advisable to determine that the *Bam*HI RE used to make cloning vectors can generate a high frequency of ligatable ends. A plasmid such as pBR322 can be cleaved by *Bam*HI to completion, followed by religation of *Bam*HI-cut plasmid with DNA ligase. Agarose gel electrophoresis of the cut and religated pieces should show that over 95% of DNA staining in the religated sample lane is in a ligated form. Otherwise, the *Bam*HI sample may not give good results.

Sucrose Gradient Resolution of Ligated Arms (32.8 kb) from Excised Piece (7.4 kb)

16. Separate arms from insert in six sucrose gradient tubes, as described in Section 13-2, loading 80 μl (one-sixth of mix from step 15) onto each gradient.
17. Centrifuge, collect drops, and analyze gradients, as described (Section 13-2). Pool fractions containing the 32.8-kb arms. The first peak to emerge from bottom of tube will be the "arms." This can be confirmed by running aliquots of the gradient fractions on an agarose gel (Section 5-5). Carefully avoid contamination with fractions containing the 7.4-kb center fragment.
18. Add 2.5 volumes of ethanol. Chill at −20°C for several hours or overnight.

Next Day

19. Spin at 12,000 × *g* for 10 min at 4°C. Decant supernatant.
20. Wash pellet with 1 ml of ethanol. Centrifuge, decant supernatant, and air-dry pellet.
21. Resuspend DNA at 0.5 μg/μl in TE buffer. Store frozen at −20°C. Check arms for purity and concentration of DNA by agarose gel electrophoresis and ethidium bromide staining (Section 9-1). Compare to a known DNA standard. Proceed to Section 13-4.

Preparation of EMBL3 Arms

EMBL3 DNA is digested with *Bam*HI to release the central stuffer fragment. Subsequent digestion with *Eco*RI removes the *Bam*HI termini from the stuffer so that it does not compete with the partial *Mbo*I insert in ligation to vector arms. Isopropanol precipitation selectively removes the *Bam*HI termini excised by *Eco*RI digestion.

Digestion of EMBL3 DNA with *Bam*HI

1. Add together:

EMBL3 DNA (500 μg)	500 μl
10× *Bam*HI buffer	100 μl
*Bam*HI (10 U/μl)	100 μl
H_2O	300 μl

2. Incubate at 37°C for 2 hr.
3. Extract with chloroform and precipitate, as in steps 3 to 9, for Charon 28 preparation.
4. Resuspend pellet in 500 μl of TE buffer.

Digestion of EMBL3 DNA with *EcoRI*

5. Add together:

DNA from step 4	500 μl
10× *Eco*RI buffer	100 μl
*Eco*RI (1,000 U)	100 μl
H_2O	300 μl

6. Incubate for 4 hr at 37°C.

7. Extract, precipitate and recover, as in steps 3–9 for Charon 28 preparation.

8. Resuspend pellet in 500 μl of TE buffer.

Isopropanol Precipitation

9. To 500 μl of DNA from step 8, add:

75 μl of 3 M sodium acetate, pH 6.0
300 μl of isopropanol in a 1.5-ml microfuge tube

10. Incubate for 15 min on ice.

11. Spin in microcentrifuge for 10 min at 4°C.

12. Decant liquid. Wash pellet with 200 μl of 0.35 M sodium acetate:ethanol (1:2.5).

13. Spin in microcentrifuge for 10 min at 4°C.

14. Repeat steps 12 and 13.

15. Decant liquid. Dry pellet under vacuum.

16. Redissolve arms at 0.5 μg/μl in TE buffer.[2]

17. Store arms at −20°C. Proceed to Section 13-4.

REFERENCES

Blattner, F., et al., Science *196:*161, 1977.
Frischauf, A.-M., Lehrach, H., Poustka, A., and Murray, N., J. Mol. Biol. *170:*827, 1983.
"Molecular Biologicals," Promega Biotec, 1984.

---

[2] It is important to check recovery of arms. To do this, an aliquot of sample is heated to 60°C to dissolve cohesive ends and run on an agarose gel in comparison to a known quantity of *Hin*dIII-cut λ DNA also heated to 60°C. The concentration of arms is estimated by comparing EB staining intensity with the intensity from the known DNA standard.

SECTION **13-4.**

Ligation of Genomic DNA into Bacteriophage Arms and Packaging to Form Library

DESCRIPTION

The prepared partial *Mbo*I-digested genomic DNA is ligated into the prepared bacteriophage arms. The best ratio of genomic DNA fragments to cloning vector for optimal library production is determined empirically in a series of trial reactions. The remainder of the genomic DNA fragments are ligated to the vector at the optimal ratio, and the recombinant phage are packaged in vitro. The packaged phage are then used to infect an appropriate *E. coli* host (such as LE392) to generate the library.

TIME REQUIRED

Day 1—1 hr
Day 2—4 hr

REAGENTS

10× ligase buffer
T4 DNA ligase, 100 U/μl (NEBL)
TMG buffer (Tris, $MgSO_4$, gelatin)
Packaging extract (e.g., Packagene from Promega Biotec)

METHODS

In Advance

Prepare partial *Mbo*I-digested insert (Section 13-2). Prepare vector arms for insertion (Section 13-3).

Ligation[1]

1. To obtain the maximal efficiency in generating library members, the optimal ratio of insert to vector must be determined empirically in a small series of test ligations and packaging reactions. The following protocol can be used: In five 1.5-ml microfuge tubes (labeled A to E), add the following:

10× ligase buffer	1 μl
Arms (0.5 μg/μl)	2 μl

2. Add to tubes

	A	B	C	D	E
*Mbo*I insert (0.25 μg/μl)	—	1 μl	2 μl	3 μl	4 μl
H_2O	6 μl	5 μl	4 μl	3 μl	2 μl

Add 1 μl of T4 DNA ligase to each tube.

There is a total of 10 μl in each tube.

3. Mix by briefly centrifuging liquid to bottom of tube. Incubate at 14°C for 12–16 hr. Ligations can be stored at −20°C until ready to package.

Packaging

4. For simplicity and convenience, a commercially available packaging extract (e.g., Packagene by Promega) is used for this step. The entire 10-μl ligation in all tubes is packaged. Follow manufacturer's directions for packaging. This involves mixing the ligation product with a sample of the packaging extract and incubating this mix for 2 hr at 22°C. If you choose to make packaging extracts, previously published methods can be followed. See the reference below.
5. After packaging, the mix is diluted to 250 μl with TMG buffer. Packaged DNA is stored at 4°C and is never frozen.
6. Aliquots of these dilutions will be plated to determine the optimal ratio for library generation (Section 13-5). A number of ligation and packaging reactions will be performed at the optimal ratio and pooled. The library generated will be titered and plated for screening and/or amplification (Section 13-5 to 13-7).

REFERENCE

Maniatis, T., Fritsch, E. F., and Sambrook, J., *Molecular Cloning: A Laboratory Manual.* Cold Spring Harbor Laboratory, New York, 1982.

---

[1] Make sure that arms and *Mbo*I insert are diluted in EDTA concentrations of less than 1 mM; otherwise ligation and packaging may not work well.

SECTION

13-5. Titering and Plating of Packaged Library

DESCRIPTION

This method allows assessment of optimal ligation and packaging conditions for generating clones. Multiple dilutions of the packaged ligation products are titered to determine optimal conditions for preparative library generation. After preparative library formation, phage may be plated for screening or the library may be amplified.

TIME REQUIRED

Day 1—2 hr
Day 2—2 hr

REAGENTS

LB medium
Maltose (20% wt/vol, sterilized by autoclaving)
10 mM $MgSO_4$
LB top agar, supplemented with **10 mM $MgSO_4$**
LB agar plates
LE392 cells
LB top agarose
  0.8 g of agarose added to 100 ml of LB medium
  Autoclave, cool to 48°C, and add 1 ml of sterile 1 M $MgSO_4$.

METHODS

In Advance

Ligate and package genomic DNA at multiple vector:insert ratios, as in Section 13-4.

Prepare Plating Cells

1. Inoculate 100 ml of sterile LB medium containing 2 ml of maltose solution with a fresh colony of LE392 bacteria in a sterile 500-ml flask.[1] Grow until O.D. at 600 nm is approximately 1.0.
2. Collect cells by centrifugation at 2,500 × *g* for 10 min in two sterile 50-ml screw-cap polypropylene tubes.
3. Decant supernatant. Resuspend each pellet in 25 ml of sterile 10 mM $MgSO_4$. This yields approximately 1–2 × 10^9 cells per milliliter.
4. LE392 cells can now be stored at 4°C for up to 2 weeks. A small decline in plating efficiency may be noted over time in storage.

Plating Phage λ (or Packaged DNA) to Determine Titer

5. Add 1 μl of each 250 μl packaging reaction (from Section 13-4 step 5) to 99 μl of TMG. Dispense 1 and 10 μl of this dilution into sterile 13 × 100 mm tubes. Be sure to also dilute and plate the cloning vector, ligated and packaged without insert (tube A from Section 13-4) as a background control.
6. To each tube, add 200 μl of LE392 bacteria from step 4. Mix. Incubate for 20 min at room temperature.
7. Add 2.5 ml of sterile LB top agar, supplemented with 10 mM $MgSO_4$, at 48°C. Pour each tube *immediately* onto a 90-mm LB agar plate at room temperature (Section 20-6). Swirl to distribute top agar evenly.
8. Allow top agar to harden for 5 min at room temperature. Invert plates and incubate at 37°C for 8–16 hr. Plaques will become evident as open areas due to lysed bacteria in the confluent lawn of LE392 cells.
9. Compute the titer of each packaging reaction. Count number of plaques per dish and compare to minus insert background control (see step 5). The addition of an insert at an optimal concentration will yield at least 5- to 10-fold the titer of the control without insert if library formation is successful.

Scaling Up

10. If a scale-up is required, repeat ligation and packaging reaction under optimal conditions, using enough samples to yield a library. Do not scale up reaction size.[2] For mammalian genomes, a reasonable target library size is 1–2 × 10^6 plaques. Pool packaging mixes and titer this pool by plating to determine the number of phage produced in the library, or the base of the library.

---

[1] LB medium, $MgSO_4$, agar, and so on must be sterile for use. Maltose will increase the λ phage receptor number in the host bacteria.

[2] For example, if the optimal ligation yielded 2 × 10^5 plaques per tube, set up 10 such ligations and reactions to give 2 × 10^6 plaques when pooled.

11. The next seven steps describe the plating of the library for screening. To amplify the library before screening, if desired, skip ahead to Section 13-7 and return to step 12 after amplification.

Plating Library for Screening

12. Generally, about 5×10^5 to 10^6 plaques are screened by plating approximately 10^4 plaques on 90-mm LB agar plates (50–100 total plates).[3] To begin, place 20 ml of LE392 cells for plating (step 4; in 10 mM $MgSO_4$) into a sterile plastic 50-ml capped tube. Add 5×10^5 to 10^6 plaques from titered packaging mixes (or previously amplified and titered library) to the bacteria. Mix.
13. Allow adsorption of phage to cells for 20 min at room temperature.
14. Pipette 200-μl aliquots of adsorbed bacteria into 50-100 sterile 13 × 100 mm glass tubes.
15. Add 2.5 ml of melted sterile LB top agarose at 48°C to each tube and pour *immediately* over an LB agar plate (at room temperature) and swirl for even distribution. Continue until all tubes are completed.
16. Allow top agarose to harden for 15 min at room temperature. Invert plates and incubate for 8–16 hr at 37°C.
17. Plaques will begin to form during this time.[4]
18. Store plates at 4°C until screening.

---

[3] Alternatively, 150 mm plates can be used with 2×10^4 plaques per plate. If these larger plates are used, then larger tubes will be used in step 14 and 7 ml volume will be used in step 15.

[4] For optimal screening, plaques should be evident but small enough not to produce confluence of lysis on the plate. To stop plaque size expansion at the optimal time, watch carefully after 8 hr and move plates to 4°C when plaques begin to appear. Plaques will continue to grow for a short while at 4°C.

SECTION **13-6.**

Screening a Plated Library with Radiolabeled Probes

DESCRIPTION

This method describes a procedure for selecting bacteriophage clones by hybridization to a radiolabeled probe. DNA from plated plaques are partially transferred to circular NC filters for hybridization with the labeled probe (from nick translation in Section 7-1 or ^{32}P end labeling of synthetic probes in Section 6-2). Several rounds of plating and screening will yield plaque-purified clones for further growth and characterization.

TIME REQUIRED

1 week

REAGENTS

0.2 M NaOH with 1.5 M NaCl
2× SSC with 0.4 M Tris, pH 7.4
2× SSC
Hybridization buffer-N or -S
Salmon sperm DNA, 2 mg/ml
10 M NaOH
2 M Tris, pH 7.4
1 M HCl
LB top agar with 10 mM $MgSO_4$
TMG buffer

METHODS

In Advance

Plate library (or amplified library) for screening at optimal density (see Section 13-5).

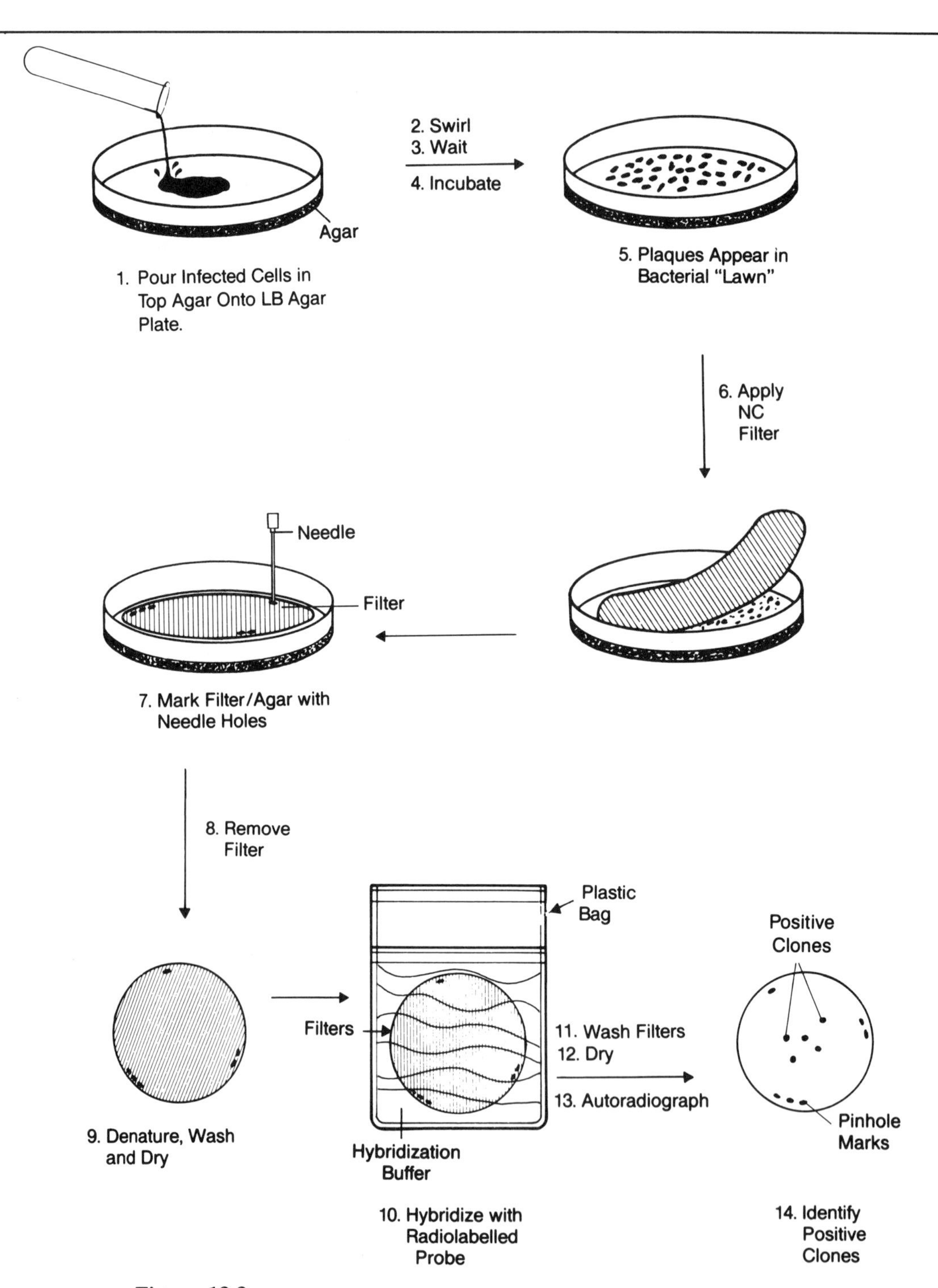

Figure 13.2
Screening of genomic library for positive clones. Following plating and growth to an optimal density, plaques are lifted to a NC or nylon filter and the DNA from the filter is hybridized with a radiolabelled probe. Location of plaques producing DNA of interest is determined by autoradiography of filters. Note that it is important to mark the corresponding positions in the plate, filter and autoradiogram to be able to reidentify the position of positive plaques.

Transfer to Filters

1. Turn plates agar side down and remove plate covers. Air-dry for about 30 min at room temperature.
2. Label each plate with a number. Label corresponding NC filter (e.g., S & S BA85, 82 mm) for each plate with ink (e.g., Skilcraft pen).
3. Carefully place filters, numbered side up, on the agar. The filter will wet in 0.5 to 1 min. Allow plaque adsorption to NC for about 20 min.
4. Punch about 5 to 10 holes through the filter and agarose with a clean 18-gauge needle (asymmetrically around outer edge) to key filter location to plate. (This is very important.)
5. Carefully and slowly peel filter off plate with flat-faced forceps. Do not dislodge top agarose layer of plate containing plaques. Mark needle holes on bottom of plate with a marking pen.
6. Dry filters at room temperature with numbered side down for about 30 min. A second set of lifts can be made for each plate. This second set is useful for eliminating false-positive signals, since all true positives should hybridize in the same location on both duplicates. Alternatively, the second set of filters may be hybridized with a second probe.
7. Dip filters sequentially (30–60 sec for each dip) in 100 ml of the following three solutions:

   a. 0.2 M NaOH, 1.5 M NaCl
   b. 2× SSC, 0.4 M Tris, pH 7.4
   c. 2× SSC

   Change solutions after every 25 filters.
8. Dry NC filter, plaque side up, for 1 hr at room temperature on Whatman 3MM paper.
9. Dry in vacuum oven for 2 hr at 80°C.

Hybridization to Filters

10. Wet all filters in hybridization buffer. Use buffer N for nick-translated or S for synthetic probes.
11. Add filters and sufficient hybridization buffer to cover filters (about 250 ml per 100 filters) in a 1-liter beaker.
12. Prepare hybridization probe. If using a synthetic probe add 10^6 cpm/ml and proceed to step 14. Transfer 0.25 μg of a nick-translated probe with a specific activity of about 10^8 cpm/μg to a 50-ml screw-cap tube. Add 0.5 ml of a 2 mg/ml solution of salmon sperm DNA. To denature add sequentially, while mixing with vortex:

    50 μl of 10 M NaOH
    300 μl of 2 M Tris, pH 7.4
    500 μl of 1 M HCl (dropwise on side of tube)

13. Add this denatured probe to the beaker with the filters and buffer. Mix by swirling.
14. Incubate with shaking (gentle agitation) at 42°C for 4–16 hr to allow hybridization of nick-translated probe. Incubation temperature for synthetic probes may vary, see Section 6-3.

Wash Filters

15. Decant hybridization buffer from beaker.[1] Wash filters by adding 500 ml of 2× SSC with 0.1% SDS (wt/vol). Shake gently for 15 min at room temperature. If synthetic probes are used, wash as described in Section 6-3, and proceed to step 18.
16. Decant and wash two more times, as in step 15. Blot all filters with paper towels before proceeding to the low salt washes.
17. Wash filters two more times in 500 ml of 0.1× SSC with 0.1% SDS at 52°C for 15 min.[2]
18. Remove wash buffer. Blot filters with Whatman 3MM paper and air dry for 1 hr.
19. Tape NC filters to sheets of 3MM paper. Mark 3MM paper with radioactive ink. Cover mounted filters with clear plastic wrap. Expose to XAR-5 film with intensifying screens (e.g., Du Pont Cronex) at −70°C overnight to detect putative positive signals.[3]

Secondary Screen

20. Aliquot 200 μl of LE392 indicator cells (see step 4 of Section 13-5) into a 13 × 100 mm tube. Add 2.5 ml of LB top agar with 10 mM $MgSO_4$ at 48°C. Pour immediately into a 90-mm LB agar plate at room temperature. Allow to cool for 5 min at room temperature, forming a "lawn" of bacterial cells. Mark a grid on bottom of plate with approximately 50 places for secondary plaques.[4]
21. Align film with radioactive ink markers over filters. Mark the positions of the filters' positioning holes on film with a marking pen. Use a transilluminator to help with alignment. Align plates with positive clones over X-ray film with positive clones (using "keys" marked on film and plates).[5]

---

[1] Handle and dispose of radioactive material safely.

[2] If imperfect homology exists between the probe and clones to be screened, the stringency of the washes may have to be lowered (e.g., lower wash temperature). Also, bath hybridization and wash conditions are different for oligonucleotide probes and will have to be adjusted individually and optimized empirically for the oligonucleotide probe used.

[3] Mark holes in NC filters with radioactive ink to allow realignment of X-ray film.

[4] Plates can be purchased with preformed grid.

[5] A transilluminating light box will help with alignment.

22. Touch a toothpick to each plaque over or near a positive autoradiographic signal on the library plates (use up to 20 toothpicks to cover a positive area).
23. Transfer plaques on toothpicks to individual squares on grid of bacterial lawn by gently puncturing lawn with end of toothpick.
24. Incubate plate (inverted) overnight at 37°C.
25. Lift to NC filter and screen, as described in steps 1 to 19.

Tertiary Screen

26. Transfer phage from a single hybridizing secondary plaque generated by each primary signal with a toothpick for each positive primary signal. Approximately 10^6 to 10^7 phage particles will be transferred with each toothpick.
27. Drop toothpick into 10 ml of sterile TMG buffer in a sterile tube. Vortex mixture to disperse phage.
28. Place 0.25 and 2.5 μl of TMG mixture from step 27 into two sterile 13 × 100 mm tubes containing 200 μl of LE392 plating cells (Section 13-5, step 4).[6] Add 2.5 ml of LB top agar with 10 mM $MgSO_4$ at 48°C. Immediately pour over LB agar 90-mm plate at room temperature. Swirl to distribute evenly. Allow to harden for 15 min at room temperature.
29. Invert plates and incubate plates overnight at 37°C. This should lead to formation of 50–500 individual discrete plaques per plate.
30. Using clean toothpicks, transfer phage from 100 to 200 well-separated plaques onto plates with a LE392 cell lawn (prepared in a grid pattern as in step 20 for secondary screen). Repeat for each tertiary screen clone being analyzed.
31. Invert plates and incubate overnight at 37°C.
32. Hybridize and screen tertiary grids, as in steps 1 to 19. Positive signals are plaque-purified clones. One representative from each primary positive "winner" is now toothpick transferred to 5 ml of TMG buffer, as described in step 26.
33. Add 1 μl of TMG mix to 200 μl of LE392 plating cells (Section 13-5, step 4) and plate, as in steps 28 and 29. Incubate inverted plates overnight at 37°C.
34. All plaques on this plate are plaque purified representations of the clone. This plate can be sealed with Parafilm and stored at 4°C as a stock of the clone. An individual plaque from this plate can be used to grow this bacteriophage preparatively (Sections 12-1 and 12-2).

REFERENCE

Benton, W., and Davis, R., Science *196:*180, 1977.

[6] If pipettor will not dispense 0.25 μl, make a 1:10 dilution into TMG buffer and dispense 2.5 μl.

SECTION

13-7. Library Amplification

DESCRIPTION

This method is used to amplify genomic or cDNA libraries in an appropriate host bacterial strain. Amplification usually yields over 10^4 copies of the library, thus preserving the library for future screening. However, amplified libraries can become skewed if some bacteriophage clones grow more efficiently during amplification than others. Thus it may be advantageous to screen a portion of the library before amplification (Section 13-6).

TIME REQUIRED

Day 1—0.5 hr

Day 2—1 hr

REAGENTS

LB soft agar

Add 1.5 g of agar to 500 ml *LB medium*.

Autoclave. Cool to 48°C.

Add 5 ml of sterile *1 M* $MgSO_4$.

NaCl (crystalline)

Chloroform

LB agar plates

METHODS

In Advance

Prepare LE392 cells for plating (Section 13-5, steps 1 to 4). Prepare and titer a packaging mix containing the library to be amplified (Sections 13-1 to 13-5).

Procedure

1. For each 2×10^4 library members, mix 200 μl of LE392 cells for plating with titered packaging mix containing about 2×10^4 phage in a sterile 13×100 mm tube. Typically, 100 tubes will be needed to amplify an entire library of about 2×10^6 plaques. Allow adsorption by incubating for 20 min at room temperature.
2. Add 2.5 ml of LB soft agar with 10 mM $MgSO_4$ at 48°C to each tube and immediately pour over a 90-mm LB agar plate at room temperature. Swirl to spread evenly. Do not invert plate or agar will slide off. Allow to grow agar side down at 37°C for 6–10 hr.
3. When phage begin to form pinpoint plaques, but before plaques enlarge to become confluent, add 2 ml LB medium and scrape soft agar with a sterile plastic scraper or bent glass pipette from all plates into a sterile 1-liter beaker. Add 2 ml of chloroform for each 100-ml volume scraped from plates.
4. Centrifuge mixture at $3{,}000 \times g$ for 10 min in sterile conical tubes. Agar and bacterial debris will form a pellet at the bottom of the tube. Decant supernatant containing amplified library.
5. Measure volume of library. Add sufficient crystalline NaCl to bring NaCl concentration to 1 M.
6. Titer amplified library by dilution and plating (Section 20-7). Typical titers are 5×10^9 to 5×10^{10} phage per milliter.
7. Store library in 1-ml aliquots in sterile tubes with 3 drops of chloroform at 4°C.
8. Retiter library before each plating and screening.

SECTION

14

cDNA Cloning into λgt10 and λgt11

SECTION

14-1. Preparation of λgt10 and λgt11 cDNA Cloning Vectors

cDNA clones differ from genomic DNA clones in that the former represent a permanent DNA copy of an mRNA. A cDNA clone is therefore representative of those parts of a gene that are expressed as RNA, in the simplest cases without introns and usually without the regulatory sequences found in genomic DNA. From a practical viewpoint, whereas genomic DNA libraries can be created by ligating appropriately digested genomic DNA in a prepared vector, cDNA libraries must be created by a series of enzymatic reactions in which a "first strand," or DNA strand complementary to RNA, is generated, followed by removal of the RNA strand and replacement by a "second strand" of DNA. Following these steps, the now double-stranded DNA is ligated into an appropriate vector. Whereas genomic DNA-cloning vectors generally accept 4–20 kb, cDNA cloning vectors will not accept more than 6–8 kb.

λgt10 and λgt11 are bacteriophage cloning vectors designed by Richard Young and Ronald Davis to generate cDNA libraries from eukaryotic mRNA. They are quite efficient in generating many clones from small amounts of DNA (i.e., about 5,000 clones per nanogram of cDNA). These vectors also have an advantage over pBR plasmids for library formation, because screening and manipulating plaques offers many technical advantages compared to screening bacterial colonies.

The vector λgt10 has a strong biological selection against lytic growth of non-cDNA-bearing bacteriophage in an appropriate host. Insertion of cDNA into λgt10 inactivates the phage repressor gene (*cI*). When λgt10 is plated on an *hfl*A (high-frequency lysogeny) mutant of c600, only phage-bearing cDNA inserts can form lytic plaques, while both insert-bearing and non-insert-bearing phage form plaques on a non-*hfl* c600 host.

λgt11 lacks this selection, but is a protein expression cDNA cloning vector. cDNA inserts in λgt11 are cloned into the carboxyl end of the β-galactosidase gene-coding region (*lac Z*). The translation products of cDNA inserts may therefore be expressed as a β-gal-cDNA fusion protein. One of every six recombinants derived from a given type of mRNA would be expected to be in the correct reading frame and orientation for appropriate translation. When the *lac Z* gene is induced with isopropyl-β-D-thiogalactopyranoside (IPTG), this fusion protein is expressed, allowing detection with antibody probes. In addition, cDNA insertion inactivates the *lac Z* gene product. Plaques formed by *lac* Z^- λgt11 bacteriophage on a *lac* Z^- host cannot metabolize 5-bromo-4-chloro-3-indolyl-β-D-galactoside (Xgal) and form clear plaques. *Lac* Z^+ nonrecombinants (noninsert containing λgt11) are gal^+ and can metabolize Xgal, producing blue plaques. This allows for colorimetric selection by distinguishing between insert-bearing (clear) and non-insert-bearing (blue) λgt11 bacteriophage plaques. λgt11 is grown with a Y1090 *E. coli* host.

TIME REQUIRED

About 2 weeks for all steps from RNA preparation to starting library screening

REAGENTS

E. coli c600 and *E. coli* c600 *hfl*A are used for λgt10 and Y1090 and BNN97 are used for λgt11. [Y1090 and BNN97 (ATCC number 37194 containing λgt11 in its genome) are available from the American Type Culture Collection]. Strains and vectors are also available commercially from Vector Cloning Systems and Promega Biotec.

Supplemented LB: *LB medium* with the following added sequentially from sterile filtered stock solutions to yield these final concentrations:

2 mM $MgSO_4$

4 μM $FeSO_4$

0.1 mM $CaCl_2$

0.15% glucose

10× *Eco*RI buffer

700 mM Tris, pH 7.4

500 mM NaCl

50 mM $MgCl_2$

Spermidine, 400 mM, pH 7.0. Store aliquots at −20°C.

*Eco*RI *(NEBL or BRL)*

SS-phenol

Chloroform

5 M NaCl

Ethanol

TE buffer

100 mM Tris, pH 8.0, with **1 mM $MgCl_2$**

Ampicillin, 50 mg/ml stock solution (store in aliquots at −20°C)

Reconstituted LCIP

Lyophilized calf intestinal phosphatase (BM)

Reconstitute at 0.6 U/μl in:

100 mM Tris, pH 8.0

10 mM $MgCl_2$

0.1 mM $ZnCl_2$ (or 0.1 mM $ZnSO_4$)

Add an equal volume of glycerol (to 50%). Store in aliquots at −20°C.

METHODS

Production of λgt10 Cloning Vector

λgt10 cloning arms are commercially available. If cloning vector is not purchased, prepare it as in steps 1 to 17. Otherwise, go to Section 14-2. λgt10 is cI^+, and has proven difficult to grow to high titer in liquid medium. This method of growth has worked well for preparing DNA from this bacteriophage.

Growth of High-Titer Lysate

1. Grow *E. coli* c600 to an O.D. at 600 nm of 0.1–0.2, at 37°C, with shaking, in supplemented LB medium.
2. Inoculate 10 ml of c600 growth with a single plaque of λgt10 punched out of a stock plate (prepared as described in Section 20-7, using c600 host cell) with a sterile Pasteur pipette.
3. Shake vigorously for 4–6 hr at 37°C until lysis is apparent (bacterial cell debris, partial clearing of solution, and foam in medium). Add 2 drops of chloroform and shake for 5 min at 37°C. Centrifuge bacterial debris at 1,500 × *g* and decant the supernatant. Store supernatant in the refrigerator at 4°C. This is a bacteriophage lysate.
4. Titer phage stock (approximately 5×10^9 to 10^{10} pfu/ml) by serial dilution and plating, using c600 as a host, as described in Section 20-7. Plaques should be turbid in the center, indicating cI^+ genotype.
5. Grow 1 liter of *E. coli* c600 in supplemented LB medium to an O.D. of about 0.2 at 600 nm.
6. Inoculate cells with 2×10^9 pfu of stock (step 4). Shake vigorously for 4–6 hr to lyse.
7. Titer growth by serial dilution and plating. There should be 3×10^9 to 10^{10} pfu/ml.
8. Concentrate phage and prepare phage DNA as described in Section 12-2, steps 7 to 31.

Preparation of Arms for Cloning: *Eco*RI Digestion

9. Pipette 100 μg of λgt10 DNA in 380 μl of TE buffer in a 1.5-ml microfuge tube.

10. Add:

    45 μl of 10× *Eco*RI buffer
    4.5 μl of 400 mM spermidine, pH 7.0
    250 U of *Eco*RI

11. Digest for 2–3 hr at 37°C. It is important to get complete digestion for efficient cDNA cloning.
12. Extract with 450 μl of a 1:1 mixture of SS-phenol and chloroform. Mix well. Spin in microfuge for 2 min. Remove upper aqueous phase to new tube.
13. Add 450 μl of chloroform to aqueous phase, mix, and spin as in step 12. Place upper aqueous phase in a new tube.
14. Add 50 μl of 5 M NaCl.
15. Add 1 ml of ethanol. Freeze tube on dry ice for 10 min. Spin in microcentrifuge for 2 min at 4°C.
16. Decant supernatant. Rinse pellet with 500 μl of 80% ethanol. Spin in microcentrifuge for 1 min.
17. Decant ethanol. Air-dry pellet. Resuspend λgt10 arms in 400 μl of TE buffer. Store at −20°C until used. Note: Phage DNA may be difficult to resuspend after ethanol precipitation. It may be necessary to leave DNA in TE buffer overnight at 4°C to achieve total resuspension. Proceed to Section 14-2.

Preparation of λgt11 Cloning Vector

Note: High-quality λgt11 dephosphorylated *Eco*RI arms are commercially available (Promega Biotec, part of Protoclone GT System). If you choose to make your own dephosphorylated arms, follow this procedure for growing λgt11. Because λgt11 contains the cI857 temperature-sensitive repressor, the simplest way to prepare bacteriophage is to induce the λgt11 lysogen BNN97 with a temperature shift to 43°C, inactivating the repressor, lysing the *E. coli* host with chloroform, to release bacteriophage. If arms are purchased, proceed to Section 14-2.

Preparation of λgt11 Arms

1. Generate single colonies of BNN97 by streaking out strain on an LB agar plate. Grow plate at 32°C.
2. Check temperature sensitivity of lysogen by streaking several single colonies onto two LB agar plates. Grow one plate at 32°C and the other at 43°C. A good temperature-sensitive lysogen should grow only at 32°C.
3. Inoculate 10 ml of LB medium with an established temperature-sensitive colony. Grow overnight to saturation with shaking at 32°C.
4. Inoculate 1 liter of LB medium containing 10 mM $MgSO_4$ with 10 ml of overnight growth from step 3. Grow with vigorous shaking at 32°C to an O.D. of 0.5–0.7 at 600 nM.

5. Induce lysogen by rapidly elevating temperature of bacteria to 43°C by shaking flask in a hot water bath. Shake vigorously at 43°C for 15 min. Be careful to maintain a steady temperature.
6. Shake culture vigorously for 3 hr at 38°C, being careful to maintain temperature high enough for induction (i.e., over 37°C).
7. Lyse the bacteria by adding 10 ml of chloroform to culture. Shake for 5 min at 37°C.
8. Add 120 ml of 5 M NaCl.
9. Prepare bacteriophage and DNA, as described in Section 12-2, steps 9 to 31.
10. Digest prepared DNA with *Eco*RI and purify, as in steps 9 to 16, for λgt10. Resuspend digested DNA in 400 μl of 100 mM Tris, pH 8, with 1 mM $MgCl_2$.

Dephosphorylate λgt11, *Eco*RI Arms

11. Add 10 μl of reconstituted LCIP to 100 μg *Eco*RI-cut λgt11, resuspended in buffer (from step 10) in a 1.5-ml microfuge tube.
12. Incubate for 1 hr at 37°C.
13. Add 400 μl of a 1:1 mixture of SS-phenol and chloroform. Mix well. Spin in microcentrifuge for 5 min.
14. Remove upper aqueous phase to a new tube. Add 400 μl of chloroform. Mix. Spin in microcentrifuge for 1 min.
15. Transfer upper aqueous phase to a new tube. Add 50 μl of 5 M NaCl.
16. Add 1 ml of ethanol. Mix well. Freeze on dry ice for 10 min. Spin in microcentrifuge for 2 min at 4°C.
17. Decant supernatant.
18. Rinse pellet with 500 μl of 80% ethanol.
19. Spin in microcentrifuge for 1 min.
20. Discard ethanol. Air-dry pellet.
21. Resuspend dephosphorylated λgt11 DNA in 400 μl of TE buffer. Phage DNA may be difficult to resuspend. Sometimes it is necessary to let pellet dissolve overnight at 4°C in TE before storage.
22. Store at −20°C until ready for use.

REFERENCES

Huynh, T., Young, R., and Davis, R., Constructing and screening cDNA libraries in λgt10 and λgt11. In: *DNA Cloning: A Practical Approach* (D. Glover, ed.), IRL Press, Oxford, 1984.

Young, R., and Davis, R., Proc. Natl. Acad. Sci., USA *80*:1194, 1983.

SECTION **14-2.**

Generation of cDNA Insert from Eukaryotic mRNA

DESCRIPTION

mRNA is isolated from eukaryotic cells and used to generate specific cDNA for insertion into λgt10 or λgt11 phage.[1] This is a long, extensive procedure for preparing an insert. The following steps are taken:

1. Isolate total RNA and select poly(A^+) RNA on oligo(dT) cellulose.
2. Concentrate poly(A^+) RNA (10 μg in 10 μl).
3. Make a complementary copy of mRNA (cDNA) by reverse transcription.
4. Synthesize the second strand by partially degrading RNA with RNase H and replacing it with DNA generated by DNA polymerase I.
5. Methylate internal *Eco*RI sites to protect the integrity of the cDNA using *Eco*RI methylase and *S*-adenosyl methionine.
6. "Polish" to blunt ends with T4 DNA polymerase.
7. Add phosphorylated *Eco*RI linkers with T4 ligase.
8. Digest excess linkers from cDNA termini with *Eco*RI to generate cohesive cloning ends.
9. Size-fractionate cDNA on acrylamide gel and electroelute.
10. Prepare cDNA insert for ligation step.

SPECIAL EQUIPMENT

Scintillation counter

[1] The method present here is adapted from that of Gubler and Hoffman (see references).

REAGENTS

0.1 M methyl mercury hydroxide (Alfa)

700 mM β-mercaptoethanol

1 M Tris-Cl, pH 8.7, autoclaved

1 M KCl, autoclaved

0.25 M $MgCl_2$, autoclaved

Oligo(dT) (12–18), 1 mg/ml, in autoclaved H_2O (P-L)

Mixture of 20 mM each dGTP, dATP, dTTP, and dCTP, pH to 7.0 (also need 2.5 mM and 1 mM mixes). Make up with autoclaved H_2O.

α-^{32}P-dCTP, 3,000 Ci/mmol (Amersham or NEN)

RNAsin, 30 U/μl (Promega)

AMV reverse transcriptase, 10 U/μl (Seikagaku America)

0.5 and 0.25 M EDTA, pH 8.0

Actinomycin D, 400 ng/ml

SS-phenol

Chloroform

4 M ammonium acetate

Ethanol and 80% ethanol

TE buffer

2 M Tris-Cl, pH 7.4 (also 10 mM)

1 M $MgCl_2$

1 M ammonium sulfate

1 M KCl

BSA, nuclease free, 10 and 5 mg/ml (Sigma, RIA grade)

RNase H, 2 U/μl

DNA polymerase I (5 U/μl) (BM)

1 M Tris-Cl, pH 8.0

50 mM EDTA, pH 8.0

S-adenosyl methionine, 150 μM (P-L)

0.15 M $MgCl_2$

*Eco*RI methylase, 20 U/μl (NEBL)

T4 DNA polymerase, 5 U/μl (BRL)

Phosphorylated *Eco*RI linkers (5′ GGAATTCC 3′), 10 O.D. units/ml in sterile H_2O (Collaborative Research)

T4 ligase, 400 U/μl (NEBL)

10 mM Tris-Cl, pH 8.0

T4 ligase buffer

- **300 mM Tris, pH 7.4**
- **100 mM $MgCl_2$**
- **100 mM DTT**
- 10 mM ATP

10× *Eco*RI buffer

700 mM Tris-Cl, pH 7.4
500 mM NaCl
50 mM $MgCl_2$

*Eco*RI (NEBL)
Gel loading buffer
5 M NaCl
5% polyacrylamide gel (Section 9-4)
TBE buffer
Size markers for gel
Yeast tRNA, 10 μg/ml
Elutip or Nensorb column

METHODS

In Advance

Prepare total RNA from tissue of interest (Section 11-1) and select poly(A^+) RNA, as described in the oligo (dT) cellulose column method (Section 11-3). dT-selected RNA is ready for cDNA preparation and library construction. The sample is stored at 1 μg poly(A^+) RNA per microliter of H_2O at −70°C.

Prepare cDNA

1. Place 10 μl poly(A^+) RNA (10 μg in H_2O) in 1.5-ml microfuge tube.[2]
2. Add 1.1 μl of 0.1 M methyl mercury hydroxide.
3. Incubate for 10 min at room temperature.
4. Add 2 μl of 700 mM β-mercaptoethanol to inactivate methyl mercury.[3] Incubate for 5 min at room temperature.
5. Add in the following order:

    5 μl of 1 M Tris-Cl, pH 8.7
    7 μl of 1 M KCl
    2 μl of 0.25 M $MgCl_2$
    10 μl of 1 mg/ml oligo(dT) (12–18)
    2.5 μl of a solution containing 20 mM each: dGTP, dATP, dTTP, dCTP,
    5 μl of α-^{32}P-dCTP

---

[2] Microfuge tubes for cDNA preparation should be sterilized by autoclaving to reduce the risk of RNase contamination of samples. Samples are relatively safe from nuclease damage after the cDNA is double-stranded.

[3] Methyl mercury is highly toxic and very volatile. Handle with extreme care and store the stock solution in a fume hood.

2 μl RNAsin, 30 U/μl
5 μl actinomycin D, 400 ng/ml
4 μl AMV reverse transcriptase, 10 U/μl

6. Incubate for 2 hr at 44°C.
7. Add 5 μl of 0.25 M EDTA, pH 8.0.
8. Add 70 μl of SS-phenol:chloroform (1:1).
9. Mix by vortexing. Separate phases by centrifugation in a microcentrifuge for 2 min.
10. Remove upper aqueous phase. Extract with 70 μl of chloroform. Repeat chloroform extraction.
11. To aqueous phase, add 60 μl of 4 M ammonium acetate. Mix well. Add 360 μl of ethanol. Mix well.
12. Freeze on dry ice for 30 min.
13. Spin in microcentrifuge for 10 min at 4°C.
14. Discard supernatant. Wash pellet with 150 μl of 80% ethanol that is kept cold but not frozen on dry ice.
15. Spin in microcentrifuge for 2 min at 4°C. Remove supernatant. Dry pellet under vacuum.
16. Resuspend pellet in 50 μl of TE buffer. Precipitate 2 μl of sample with trichloroacetic acid, as described for nick translation (Section 7-1), and count filter using a scintillation counter to determine the amount of ^{32}P-dCTP incorporated into the first strand. The efficiency of mRNA copying can be estimated at this time. In this protocol, 5×10^4 cpm of ^{32}P incorporated in the 2-μl aliquot corresponds to about 1 μg of the first strand copied. A good reverse transcription yields 5–30% of dT selected mRNA copied [i.e., 0.5–3 μg of cDNA/10 μg poly(A^+) RNA]. The average size of the first-strand cDNA synthesized can also be determined by polyacrylamide gel electrophoresis (Section 9-4).
17. Make 2× second-strand buffer:

8 μl 2 M Tris-Cl, pH 7.4
4 μl 1 M $MgCl_2$
8 μl 1 M ammonium sulfate
80 μl 1 M KCl
4 μl nuclease-free BSA (10 mg/ml)
13 μl of a mixture of 2.5 mM each dGTP, dATP, dTTP, dCTP neutralized to pH 7.0
283 μl H_2O

18. Combine:

50 μl first-strand mix (step 16)
50 μl 2× buffer (step 17)

1.0 μl RNase H (2 U/μl)
2.3 μl DNA polymerase I (5 U/μl)

19. Incubate at 12°C for 1 hr. Follow with an incubation at 22°C for 1 hr.

20. Add:

4.0 μl of 0.5 M EDTA, pH 8
100 μl of SS-phenol/chloroform (1:1 mix)

Mix well. Spin in microcentrifuge for 2 min to separate phases.

21. Transfer upper aqueous phase to new microfuge tube. Add 100 μl of chloroform. Mix well. Spin in microcentrifuge for 2 min.

22. Repeat step 21.

23. Add 100 μl of 4 M ammonium acetate to upper aqueous phase from step 22.

24. Add 600 μl of ethanol. Freeze on dry ice for 30 min.

25. Spin in microcentrifuge at 4°C for 10 min. Decant supernatant.

26. Resuspend pellet in 50 μl of H_2O.

27. Add 50 μl of 4 M ammonium acetate.

28. Add 300 μl of ethanol. Freeze on dry ice for 30 min.

29. Spin in microcentrifuge at 4°C for 10 min. Decant supernatant carefully.

30. Add 150 μl of 80% ethanol (kept cold on dry ice) to pellet.[4]

31. Centrifuge sample for 2 min at 4°C. Carefully decant supernatant, being sure not to disturb the pellet.

32. Dry pellet under vacuum.

33. Resuspend double-stranded cDNA pellet in 25 μl of autoclaved H_2O in a 1.5-ml microfuge tube.

*Eco*RI Methylation Reaction

34. To 25 μl of cDNA from step 33 add:

5 μl of 1 M Tris-Cl, pH 8
5 μl of 50 mM EDTA, pH 8
5 μl of nuclease-free BSA (4 mg/ml)
5 μl of 150 μM *S*-adenosyl methionine
5 μl of *Eco*RI methylase (20 U/μl)

35. Incubate for 20 min at 37°C.

36. Stop enzyme by incubation at 65°C for 10 min.

---

[4] The cDNA is reprecipitated with ethanol to remove nucleotides that could interfere with the *Eco*RI methylase reaction.

To "Polish" Blunt Ends of cDNA

37. To 50 μl of reaction products in step 36, add:

6 μl of mixture of dGTP, dATP, dCTP, dTTP, at 1 mM each, neutralized to pH 7.0
6 μl of 0.15 M $MgCl_2$
3 μl of T4 DNA polymerase (5 U/μl)

38. Incubate for 15 min at 37°C.

39. Add 8 μl of 0.5 M EDTA, pH 8.0.

40. Add 70 μl of a 1:1 mixture of SS-phenol and chloroform. Mix well.

41. Spin in microcentrifuge for 2 min to separate phases.

42. Remove upper aqueous phase and extract with 1 volume of chloroform. Mix. Spin as in step 41.

43. Repeat step 42. Remove upper aqueous phase to a new tube.

44. Add:

75 μl of 4 M ammonium acetate
450 μl of ethanol, kept cold on dry ice

Mix well. Freeze on dry ice for 30 min.

45. Spin in microcentrifuge at 4°C for 10 min. Decant supernatant.

46. Wash pellet with 150 μl of 80% ethanol (kept cold on dry ice). Spin in microcentrifuge for 2 min at 4°C.

47. Carefully decant supernatant. Dry pellet under vacuum.

48. Resuspend pellet in 22 μl of 10 mM Tris-Cl, pH 8.

*Eco*RI Linker Addition

49. To 22 μl of polished cDNA (from step 48) add:

8 μl of phosphorylated *Eco*RI linkers (O.D. of 10 at 260 nm in 1 ml)
4 μl of 10× T4 ligase buffer
4 μl of T4 ligase (400 U/μl)

Mix, being careful to keep mixture below 15°C because T4 ligase is heat sensitive. Note that there is a large excess of linkers compared to cDNA in order to drive the linker addition reaction.

50. Incubate overnight at 15°C. Inactivate ligase by incubation for 15 min at 65°C.

*Eco*RI Digestion to Generate Cohesive *Eco*RI Cloning Ends

51. To 38 μl of mix (from step 50) add:

10 μl H_2O
6 μl 10× *Eco*RI buffer
60 U *Eco*RI

52. Incubate digest for 2 hr at 37°C.

53. Add:

6 μl of 0.5 M EDTA, pH 8.0
60 μl of a 1:1 mixture of SS-phenol and chloroform

Mix well.

54. Spin in microcentrifuge for 2 min to separate phases. Remove upper aqueous phase to new tube.

55. Add 6 μl of 5 M NaCl and 150 μl of ethanol. Mix well.

56. Freeze on dry ice for 10 min. Spin in microcentrifuge at 4°C for 10 min.

57. Decant supernatant. Dry pellet under vacuum.

58. Redissolve pellet in 25 μl of TE buffer. Add 5 μl of gel loading buffer.

Size Fractionation of cDNA Insert

59. To remove excess *Eco*RI linkers and cDNA fragments too small to be useful, cDNA can be size fractionated on a 5% polyacrylamide gel run in TBE buffer (Section 9-4). The sample is run in parallel with size markers.[5] After 2–3 hr of electrophoresis at approximately 10 V/cm, all cDNA migrating at over 500 bp, as determined by autoradiography and comparison to size markers, is excised as a single gel piece.

---

[5] A convenient set of size markers for this gel is pBR322 digested with *Hin*fI and *Eco*RI, and end-labeled with large fragment DNA polymerase I and ^{32}P-dATP. To prepare these size standards, mix together:

1–2 μg pBR322 DNA	10 μl
10× *Eco*RI buffer	2 μl
α-^{32}P-dATP	4 μl
*Hin*fI (10 U/μl)	1 μl
*Eco*RI (10 U/μl)	1 μl
H_2O	2 μl

Incubate for 60 min at 37°C. Add 1 μl of DNA polymerase I, large fragment (5 U/μl). Incubate for 15 min at 37°C. Inactivate enzymes by heating the sample to 65°C for 10 min. The unincorporated ^{32}P-dATP can be removed by applying the labeled sample to a 5-ml G50 Sephadex column run in TE buffer. The incorporated counts can then be pooled in the void volume (first peak eluted from the column) and concentrated by ethanol precipitation (Section 20-1). Alternatively, the free label can be removed by Elutip purification of the digest (Section 10-3).

Molecular weight marker sizes: 998, 631, 517, 506, 396, 344, 298, 221, 220, 154, and 75 bp.

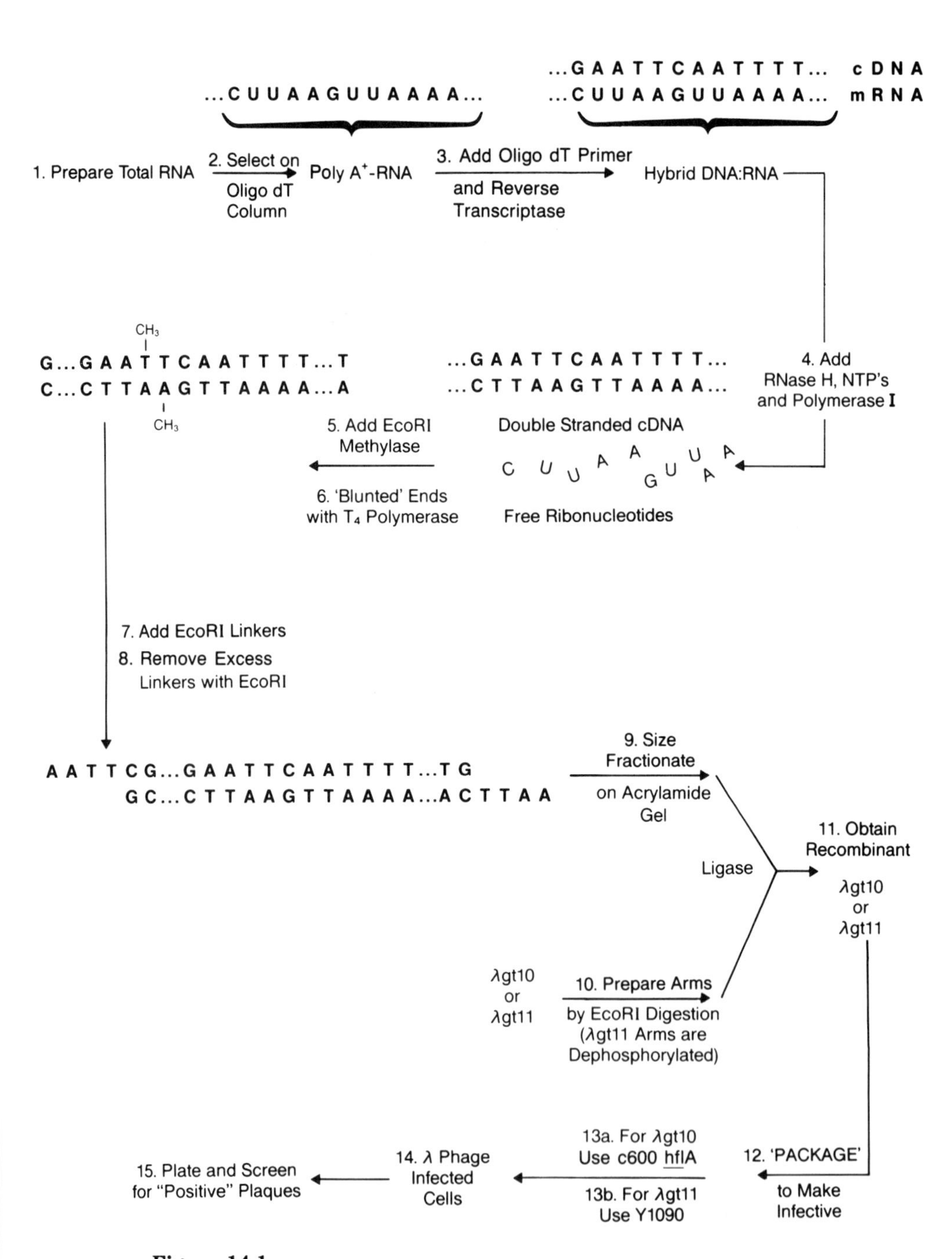

Figure 14.1
Schematic diagram of procedure for generating double-stranded cDNA species from RNA, adding *Eco*RI ends to the cDNA, and cloning into λgt10 and λgt11.

60. cDNA is eluted from the gel piece by electroelution procedure (Section 9-3). Put sample in a dialysis bag, clipped at each end, with 1 ml of 1× TBE buffer and 10 μg/ml of yeast tRNA. Elute for 12–16 hr.

61. cDNA can be concentrated from buffer in bag by passing through Elutip-d minicolumn (Section 10-3).

62. Precipitate eluate with ethanol (Section 20-1).

63. Resuspend cDNA in TE buffer at an estimated concentration of approximately 2–5 ng/μl, estimated from first-strand analysis. Estimate final yield of cDNA based on the specific activity of first strand cDNA synthesized. Store cDNA at −20°C.

REFERENCES

Gubler, U., and Hoffman, B., Gene *25*:263, 1983.

Toole, J., Knopf, J., Wozney, J., Sultzman, L., Buecker, J., Pittman, D., Kaufman, R., Brown, E., Shoemaker, C., Orr, E., Amphlett, G., Foster, B., Coe, M., Knutsen, G., Fass, D., and Hewick, R., Nature *312*:342, 1984.

SECTION

14-3. Ligation and Packaging of cDNA Library into λgt10 or λgt11 Arms

DESCRIPTION

Prepared cDNA from Section 14-2 is inserted into cloning vectors from Section 14-1. The vector and cDNA must then be packaged. A commercially available packaging kit is used.

TIME REQUIRED

5 hr for ligation

3 hr for packaging

REAGENTS

λgt10 or λgt11 arms: 0.25 μg/μl in *TE buffer* (From Section 14-1)

10× ligase buffer

300 mM Tris, pH 7.4

100 mM DTT

100 mM $MgCl_2$

10 mM ATP

Control insert

*Eco*RI-cut pBR322; digest 10 μg pBR322 DNA to completion with *Eco*RI. Purify digested DNA with SS-phenol/chloroform extraction, chloroform extraction, and ethanol precipitation (Section 20-1). Rinse pellet with 80% ethanol and dry pellet under vacuum. Resuspend so that DNA concentration is approximately 100 ng/μl in TE buffer.

T4 ligase (NEBL)
TMG buffer
Packaging extract (e.g., Packagene from Promega Biotec; Gigapack from Vector Cloning System)

METHODS

In Advance

Prepare cDNA for insertion (Section 14-2). Prepare λgt10 or λgt11 arms (Section 14-1).

Ligation

1. Add prepared λgt10 or λgt11 arms, 10× ligase buffer, model insert, cDNA insert, T4 ligase, and H_2O to six tubes, using the following suggested protocol:

	Tube					
	A	B	C	D	E	F
λgt10 or 11 arms	1 μl ⟶					
10× Ligase buffer	1 μl ⟶					
Control insert	—	1 μl	—	—	—	—
cDNA insert	—	—	1 μl	2 μl	3 μl	5 μl
H_2O	7 μl	6 μl	6 μl	5 μl	4 μl	2 μl
T4 ligase	1 μl ⟶					

There is a total volume of 10 μl in each tube.

2. Incubate tubes for 4 hr at 14°C.[1]

In Vitro Packaging

3. Commercially available packaging extracts (e.g., Packagene or Gigapack) work well for these library constructions.[2]

4. Package each ligation tube (A through F) by adding the entire ligation reaction to packaging extract (e.g., one Packagene tube).

---

1 Take care not to warm up ligation mix, because T4 ligase is temperature sensitive.

2 If you wish to prepare your own packaging extracts from *E. coli* instead of purchasing commercial extracts, a good protocol for making sonicated extracts from *E. coli* BHB2690 and freeze-thaw lysates from *E. coli* BHB2688 can be found in *Molecular Cloning: A Laboratory Manual* by T. Maniatis, E. F. Fritsch, and J. Sambrook, Cold Spring Harbor, New York, 1982.

5. In addition, package two other tubes as controls for packaging and insert-specific clones:

    G. 0.25 μg uncut λgt10 or λgt11 DNA and
    H. 0.25 μg *Eco*RI cut, unligated λgt10 or λgt11 arms (equivalent to tube B, but without T4 DNA ligase.

6. Incubate packaging reactions at 22°C for 2 hr.
7. Dilute packaging reactions to 250 μl with TMG buffer.
8. Packaging reactions can be stored at 4°C for several days without significant loss of titer.

SECTION **14-4.**

Plating and Screening of λgt10 and λgt11 Packaged Inserts

DESCRIPTION

This method is used to titer and generate an optimal cDNA library. The trial ligation and packaging reactions from Section 14-3 are plated to determine the optimal conditions for preparative library formation.

TIME REQUIRED

Several days

REAGENTS

10 mM $MgSO_4$

IPTG, 100 mM

Xgal, 10% in DMF

Top agar for λgt10

LB medium with 0.8% agar and 10 mM $MgSO_4$

Top agar for λgt11

LB medium with 0.8% agar containing:

10 mM $MgSO_4$

Ampicillin, 50 μg/ml

100 mM IPTG, 3 μl/ml

10% Xgal, 3 μl/ml

METHODS

In Advance

Prepare packaged libraries with λgt10 or λgt11 (Section 14-3). Prepare bottom plates: *LB agar* for λgt10; LB agar with ampicillin (50 μg/ml) for λgt11.

Plating

1. Plate λgt10 recombinants at appropriate dilutions (see below) on both *E. coli* c600 and c600 *hflA* plating cells prepared in 10 mM $MgSO_4$. Follow directions for plating packaging reactions found in the section on plating bacteriophage (Section 20-7). Use appropriate top and bottom agar for λgt10. Incubate plates at 37°C.

   Plate λgt11 recombinants at appropriate dilutions on *E. coli* Y1090 plating cells, grown in LB medium with 50 μg/ml ampicillin and resuspended in 10 mM $MgSO_4$. Follow directions for plating packaging reactions found in Section 20-7. Use appropriate top and bottom agar for λgt11. Incubate plates at 42°C instead of 37°C.

Suggested Plating Dilutions for the Tubes from Section 14-3

2. a. For uncut control bacteriophage DNA sample (tube G), plate 5 and 50 μl of a 1:1,000 dilution (into TMG buffer) of the packaging mix, from Section 14-3.
   b. For *Eco*RI digested, unligated control samples (tube H), plate 5 and 50 μl of a 1:10 dilution (into TMG buffer) of the packaging mix.
   c. For all other samples (tubes A through F), plate 5 and 50 μl of a 1:100 dilution (into TMG buffer) of the packaging mix.

   Note: Depending on the packaging efficiency, these dilutions may have to be adjusted to generate a readily countable number of plaques on the titer plates.

Expected Results for λgt10 Recombinants

a. On c600, both recombinant and nonrecombinant phage will form plaques. Insert-bearing λgt10 forms a clear plaque (cI^-), and non-insert-bearing λgt10 forms a turbid plaque.
b. On c600 *hfl*A, only clear plaques (cI^-) are seen, representing insert-bearing recombinants.
c. Uncut λgt10 should give about 10^8 pfu per microgram of input DNA in c600 and few to no plaques on c600 *hfl*A.
d. *Eco*RI-cut, unligated λgt10 should give less than 10^6 pfu per microgram of DNA on *E. coli* c600 plating cells. If more plaques are seen, the *Eco*RI digest generating λgt10 arms should be repeated.
e. Model insert should give 10^5 to 10^6 clear plaques per microgram of vector plated on c600, against a background of 10^6 to 10^7 turbid plaques. Only the clear plaques should be present on c600 *hfl*A.

f. *Eco*RI cut and ligated λgt10 without insert (tube A) should give few if any clear plaques on *hfl*A and 10^6 to 10^7 turbid plaques per microgram of input vector on c600. A high background of clear plaques in this sample indicates that either the *Eco*RI vector cloning site is damaged or the vector DNA (λgt10) is contaminated with other *Eco*RI-clonable DNA.
g. Addition of cDNA insert should stimulate production of clear plaques over zero insert background on c600 to an optimum value of 1 clear plaque per 10 turbid plaques. Only the clear plaques should be visible on c600 *hfl*A.

3. Choose the tube with a ratio of cDNA insert to λgt10 vector that yielded the maximum number of clear plaques over background without exceeding 1 clear plaque per 10 turbid ones on c600 to reduce the chance of generating clones with double inserts.
4. Ligate the remainder of cDNA into arms in several individual reactions identical to the trial reaction but only using the optimal ratio. Do not scale up the ligation reaction; do multiple small reactions (see Section 14-3, step 1)
5. Package each ligation reaction individually (Section 14-3, steps 3 to 8). Pool the individual reactions after dilution with TMG buffer.
6. Titer the library by serial dilution and plating, as described in Section 20-7 to determine the number of recombinant clones generated.
7. The library now may be plated for screening on c600 *hfl*A at 5×10^3 to 10^4 clear plaques per 90 mm plate. Screen for the cDNA clone of interest with a radiolabeled nucleic acid probe, as described in the section on screening genomic libraries (Section 13-6). Alternatively, the library may be amplified before screening to preserve many copies of the library for subsequent plating and screening. To amplify the λgt10 library follow the procedure in Section 13-7, but use media and *E. coli* host appropriate for λgt10 (LB with 10 mM $MgSO_4$, c600 *hfl*A).

Expected Results For λgt11

a. On Y1090, blue plaques are nonrecombinant λgt11. Clear plaques represent insert-bearing bacteriophage.
b. Uncut λgt11 (tube G) should show about 10^8 blue plaques per microgram of DNA packaged.
c. *Eco*RI cut, unligated λgt11 (tube H) should show fewer than 10^6 blue plaques per microgram of DNA. If more than this is observed, the *Eco*RI digestion was not optimal. Redigest with *Eco*RI and dephosphorylate.
d. *Eco*RI cut, dephosphorylated, and ligated λgt11 without insert (tube A) should show fewer than 10^6 blue plaques and very few clear plaques per microgram of DNA. Clear plaques indicate either damage to the *Eco*RI cloning site or DNA contamination of the λgt11 vector arms preparation.
e. Control insert (tube B) should give a significant stimulation in the number of clear plaques seen in comparison to the ligation without added insert.

Approximately 10^6 clear plaques per microgram of vector plated is usually attained. These clear plaques represent "model insert" cDNA clones.

f. Addition of cDNA (tubes C–F) should mimic the control insert (tube B) and should stimulate the numbers of clear plaques. Most of the λgt11 libraries constructed with this protocol show a 1:1 mix of clear and blue plaques as the best achievable ratio.

3. Choose the ratio of cDNA insert/λgt11 vector which yields the best ratio of clear to blue plaques (e.g. typically 1:10 to 1:1).
4. Ligate the remainder of cDNA into λgt11 arms in individual reactions at that ratio. Do not scale up the size of the ligation; use multiple tubes (Section 14-1).
5. Package each reaction individually (Section 14-3, steps 3 to 8). Pool the reactions after dilution into TMG buffer.
6. Titer the pooled library by plating an appropriate dilution to determine the number of cDNA clones generated, as described in Section 20-7.
7. The λgt11 library may be plated on Y1090 and screened with either an antibody probe or a nucleic acid probe.

Notes on Screening Libraries

1. For optimal screening of λgt10 libraries, plaque size should be kept small by using dry LB bottom agar plates (over 2 days old) and visually monitoring plaque growth. When plaques have formed (6–8 hr after plating) and are not yet confluent, move plates to 4°C to stop growth of bacteriophage plaques. The method for screening is essentially as described in Section 13-6, except that the *E. coli* host is c600 *hfl*A.
2. Nucleic acid probes can be previously cloned, nick-translated DNA or oligonucleotides. Use of nick-translated probes in screening is described in Sections 13-5 and 13-6. The use of oligonucleotide probes is described in this manual. Conditions for hybridization and wash are individually tailored, depending on the length and properties of the probe used in the experiment (Sections 6-3 and 7-2).
3. Screening of λgt11 libraries with a nucleic acid probe is performed similarly to screening a genomic library (Sections 13-5 and 13-6), except that top agarose and bottom agar contain 50 μg/ml ampicillin and the *E. coli* host is Y1090. Grow all λgt11 library plates at 42°C to inactivate phage repressor and favor lytic growth. IPTG and Xgal are not needed in the top agarose because color selection is not important in screening. The protocol for screening a genomic library is found in Section 13-6. Also, the number of plaques screened in a λgt11 library must be increased to screen a library completely; only a fraction of the plated phage bear insert.

4. Screening λgt11 libraries with an antibody probe requires an entirely different protocol for making NC lifts. Several protocols are developed for this procedure, one of which is found in the references below.[1]

REFERENCE

Huynh, T., Young, R., and Davis, R. In: *DNA Cloning: A Practical Approach* (D. Glover, ed.) IRL Press, Oxford, 1984.

Young, R., and Davis, R., Proc. Natl. Acad. Sci, USA *80:*1194, 1983.

---

[1] The antibody probe screen can be performed as follows:

a. Incubate plated λgt11 library for 2–3 hr at 42°C.

b. Overlay with numbered NC filters. Presoak filters in 10 mM IPTG and air dry before use.

c. Incubate NC-covered plates for 3–5 hr at 42°C.

d. Mark filters for alignment and carefully remove. Store plates at 4°C.

e. Wash filters three times, 20 min each, in 10 mM Tris, pH 8 with 150 mM NaCl (TBS).

f. Incubate filters for 1–2 hr in TBS with 0.2% BSA at room temperature.

g. Incubate for 1–2 hr with the titered antibody (see Section 21-2) in TBS with 1% BSA (3 ml per filter).

h. Wash three times, 20 min each, in TBS containing 0.05% Tween-20.

i. Incubate with ^{125}I-Protein A (300,000 cpm/ml) in TBS with 1% BSA for 1–2 hr.

j. Wash four times, for 20 min each, in TBS containing 0.05% Tween-20.

k. Autoradiograph filters and align film with the original plate.

l. Pick putative clones and perform additional plating and screening (Sections 13-5 and 13-6).

SECTION

14-5. Preparation of DNA from λgt10 and λgt11 cDNA Clones

DESCRIPTION

This method describes the purification of bacteriophage DNA from a 10-ml lysate of λgt10 or λgt11 grown in *E. coli* c600 *hfl*A or Y1090, respectively. A small amount (2-5 μg) of DNA is purified for preliminary analysis of cDNA clones. cDNA prepared by this method can also serve as the source of an insert for subcloning and preparative growth in a more efficient plasmid host.

TIME REQUIRED

6–8 hr

REAGENTS

For λgt10 cDNA clones: Host cell is *E. coli* c600 *hfl*A.
Medium is LB with 10 mM $MgSO_4$.
For λgt11 cDNA clones: Host cell is *E. coli* Y1090.
Medium is LB with 10 mM $MgSO_4$ and 50 μg/ml ampicillin.
Chloroform
TM buffer
50 mM Tris, pH 7.4
10 mM $MgSO_4$

DNase I, 1 mg/ml in TM buffer (Sigma #D-4263)
5 M NaCl
Polyethylene glycol 6000 (PEG-6000)
0.5 M EDTA, pH 8.0
SS-phenol

0.3 M sodium acetate, pH 8.0
Ethanol
TE buffer

METHODS

In Advance

Prepare a 2-ml overnight growth of appropriate host cells. Prepare a freshly plated plaque of cDNA clone for growth (Section 20-7). Prepare appropriate growth medium using sterile supplies.

Grow 10 ml Lysate

1. Grow host cells in appropriate medium to O.D. at 600 nm of 0.1–0.2. Do not exceed 0.2.[1]
2. Inoculate 10 ml of host cells in medium with one plaque of cDNA clone in a sterile tube (e.g., 50-ml polypropylene screw-cap). Plaque is transferred as an agar "plug" with a sterile Pasteur pipette (Section 12-1, note 1). A mock-infected bacterial culture, containing host cells but no added phage, can be grown along with the infected samples. Comparison between the infected and mock-infected samples allows easier visualization of lysis.
3. Allow culture to shake vigorously (over 200 rpm on an orbital shaker) with good aeration for 4–6 hr until lysis of bacteria is evident (medium becomes clear). Shake λgt10 clones at 37°C. Shake λgt11 clones at 42°C.
4. After lysis, add 200 μl of chloroform. Shake for 2 min at 37°C to lyse any remaining bacteria.
5. Spin sample at 3,000 × g for 10 min to remove bacterial debris. Decant supernatant into fresh tube.
6. Lysate can be stored at 4°C for a long period of time without significant loss of titer. Typical titers are 2×10^9 to 2×10^{10} pfu/ml.
7. Store 2 ml of lysate as a stock of the clone in a sterile vial, with 1 drop of chloroform, at 4°C.

Preparation of DNA from Bacteriophage

8. Add 8 ml of TM buffer to the remaining 8 ml of lysate in a 30-ml thick-walled glass (Corex) tube.
9. Add 320 μl of DNase I, 1 mg/ml in TM buffer.
10. Incubate for 15 min at room temperature.
11. Add 1.6 ml of 5 M NaCl and 1.8 g of solid PEG-6000. Vortex to dissolve PEG in lysate. Incubate for 15 min on ice.

---

1 If bacteria do not lyse (step 3), O.D. of starting host cells can be reduced to 0.05. Host cells can also be prepared in medium containing maltose (see page 183).

12. Centrifuge the precipitated phage for 10 min at 10,000 × g in a swinging bucket rotor.
13. Decant supernatant. Resuspend cloudy pellet in 300 μl of TM buffer. Transfer to 1.5-ml microfuge tube.
14. Add 300 μl of chloroform. Mix. Spin in microcentrifuge for 5 min to separate phases. Repeat this chloroform extraction.
15. Remove upper aqueous phase to a new tube. Add 15 μl of 0.5 M EDTA, pH 8, and 30 μl of 5 M NaCl, and 350 μl of SS-phenol. Mix by vortexing.
16. Separate phases by spinning in a microcentrifuge for 2 min at room temperature.
17. Remove upper aqueous phase to a new tube. Add 350 μl of chloroform. Mix by vortexing. Separate phases by microcentrifugation for 2 min.
18. Remove upper aqueous phase to a new tube. Add 875 μl of ethanol.
19. Precipitate on wet ice for 15 min. Spin out precipitate in microcentrifuge for 5 min at 4°C. Decant supernatant.
20. Resuspend pellet in 100 μl of 0.3 M sodium acetate, pH 8. Add 250 μl of ethanol. Precipitate DNA on dry ice for 10 min. Spin out precipitate in microcentrifuge for 5 min at 4°C. Carefully remove supernatant with a micropipettor to avoid disturbing pellet.
21. Rinse pellet with 150 μl of 80% ethanol. Spin in microcentrifuge for 2 min at 4°C. Carefully remove supernatant with a micropipettor without disrupting the pellet.
22. Dry pellet under vacuum. Resuspend in 100 μl of TE buffer. Store DNA at −20°C. The yield of phage DNA will be about 3 μg per 10^{10} pfu/ml initial titer.
23. cDNA sample can be now used for analysis of insert size and hybridization properties.[2] Also, cDNA inserts of interest can be subcloned into pBR322 for preparative growth.[3]

---

[2] From the analytical DNA preparation, the size of cDNA can be determined by digesting an aliquot with *Eco*RI and resolving the insert from the cloning vector by agarose gel electrophoresis, along with size markers. The small amount of insert may be difficult to visualize with ethidium bromide. Therefore, the gel should be blotted and hybridized with the original probe to determine insert size and hybridization properties (Sections 5-5, 5-6, 6-3, 7-1, and 7-2).

[3] cDNA inserts may be subcloned into calf intestinal phosphatase-treated, *Eco*RI-digested pBR322. The *Eco*RI-digested cDNA is ligated into the plasmid vector and transfected into $CaCl_2$-competent LE392 cells (Sections 15-2 to 15-4). The resulting colonies are screened with an appropriate probe in colony hybridization to determine which colonies contain insert-bearing cDNA plasmids (Section 15-3). Preparative quantities of cDNA can be easily obtained by growing up the plasmid subclones. See Sections 8-2 through 8-4 and 15-1. In addition, the cDNA inserts can be cloned directly into *Eco*RI-digested M13 vectors (mp8, mp9, mp10, mp11, mp18, and mp19) for nucleotide sequencing (Sections 16-1 to 16-10).

SECTION

15

Subcloning into Plasmids

SECTION

15-1. Subcloning into Plasmids: General Notes

DESCRIPTION

Plasmids are very useful subcloning vectors because they can be easily transfected into cells, amplified, and purified to yield large quantities of DNA. Inserts cloned into plasmids are also easier to analyze by RE mapping than those obtained from bacteriophage vectors. However, bacteriophage are more versatile for generating high-base libraries of cloned DNA. They are also easier to screen for positive clones and to amplify for repeated screening and permanent storage. For these reasons, once a positive clone is identified from a bacteriophage cDNA or a genomic library, it is recommended that sequences of interest from these clones be subcloned into a plasmid vector for further preparative growth and detailed analysis.

Some of the earlier protocols for cDNA cloning called for cDNA generated from mRNA to be cloned initially in pBR322 to form a library with clones of interest selected by colony hybridization. One frequently used protocol specified the pBR322 to be cut with the RE *Pst*I and the 3′ terminus to be extended with terminal deoxynucleotide transferase to form a single-stranded G-tail. The cDNA inserts to be cloned were synthesized to contain a complementary C-tail. The cDNA and the cloning vector were then annealed by hybridization, and the circular DNA was used to transform *E. coli* bacteria. These tailing reactions, the subsequent annealing into pBR322, and the transformation procedure are thoroughly described in detail in reference 2 below. In general it is more efficient to make large-base libraries of cDNA in λgt10 or λgt11 than in pBR322-derived plasmids. Additional advantages of bacteriophage cDNA libraries include the ease and flexibility of screening with antibody or nucleic acid probes, and a straightforward means of amplifying the library. It is recommended that plasmid vectors be used for subcloning and preparative growth of subcloned fragments, rather than for library formation.

A number of sections in this manual call for subcloning DNA into plasmids or isolating inserts previously cloned into plasmid vectors. The most commonly cited plasmid vector is pBR322, although modified versions such as pBR327 are also used. In addition, the pUC plasmids, containing a polylinker site in the *lac Z* gene are also commonly used and will be described (Section 15-4). The next section (15-2) describes the preparation of plasmid vectors for subcloning and ligation with inserts derived from a number of sources. Lastly, screening of bacterial colonies with hybridizing probes for selection of inserts of interest will be described in Section 15-3.

REFERENCES

Bolivar, F., Rodriguez, R. L., Greene, P. J., Betlach, M. C., Heynecker, H. L., and Boyer, H. W., Gene *2*:95, 1977.

Maniatis, T., Fritsch, E. F., and Sambrook, J., *Molecular Cloning: A Laboratory Manual*, Cold Spring Harbor, New York, 1982.

Sutcliffe, J. G., Proc. Natl. Acad. Sci., USA *75*:3737, 1978.

SECTION **15-2.**

Preparing pBR322 Plasmids for Subcloning and Ligation of Insert

DESCRIPTION

This section describes a method for subcloning genomic DNA fragments from genomic bacteriophage clones or cDNA inserts from λgt10 or λgt11 cDNA clones or fragments from other plasmids. pBR322 is a commonly used plasmid and serves as the example presented in this section. Other plasmids, such as pBR325 or pBR327, can be substituted with little modification of the method. The pUC family of pBR322-derived plasmids has special properties for cloning, growth, and selection, and will be described in Section 15-4.

In this method, plasmids are cut with one or more RE to a linear form. The RE chosen for vector preparation should generate termini in the vector that are compatible with the cohesive and/or blunt termini found on the insert fragment to be subcloned. For example, if an *Eco*RI-*Bam*HI fragment is being subcloned, then the vector should be digested with *Eco*RI and *Bam*HI to allow efficient formation of recombinants during ligation. Some REs generate cohesive termini that are compatible with ligation to termini from different REs.

For example, the RE *Sau*3A produces a terminus

```
▼
|G A T C
 C T A G|
        ▲
```

which is able to ligate into a *Bam*HI site

```
  ▼
G|G A T C C
C C T A G|G
          ▲
```

Any blunt terminus is compatible with ligation to another blunt terminus.

For example, a *Pvu*II terminus

```
      ▼
C A G|C T G
G T C|G A C
      ▲
```

can be ligated to a *Sma*I terminus

```
      ▼
C C C|G G G
G G G|C C C
      ▲
```

Note that ligation of termini from two different REs frequently fails to regenerate a recognition site for one or both RE. Any RE terminus, of course, can be ligated into a terminus generated by the same RE with consequent regeneration of the RE recognition site. Compatibility of RE ends in vector-insert ligation must be confirmed before proceeding with the subcloning experiment. It is also necessary that the RE-digested vector contain both an intact drug resistance gene (e.g., ampicillin or tetracycline) and an origin of DNA replication that allows selection for *E. coli* transformation with the recombinant plasmid and autonomous replication of the plasmid DNA circle in the *E. coli* host.

When a fragment with the same RE termini on both ends is being subcloned (symmetric cloning), alkaline phosphatase may be used to remove the 5′ phosphate from the linearized plasmid vector. This inhibits the compatible ends of the plasmid vector from religating to each other, enhancing the frequency of ligation to insert termini and formation of insert-bearing clones. After vector preparation, the fragments to be subcloned are ligated together with the prepared vector. This ligation mixture is transformed into *E. coli* and transformants selected using the drug resistance gene on the plasmid.

TIME REQUIRED

1 day

SPECIAL EQUIPMENT

Bacterial cell culture equipment

REAGENTS

DNA insert fragment of interest cloned into bacteriophage, (e.g., EMBL3, Charon 28, λgt10, or λgt11) or cloned into another plasmid

pBR322 plasmid or another similar subcloning vector to receive insert

RE, such as *Eco*RI

10× RE buffer (Section 5-4)

SS-phenol

Chloroform

Ethanol

TE buffer

10× ligase buffer

T4 DNA ligase, 100 U/μl (NEBL)

Lyophylized Calf Intestinal Alkaline Phosphatase (LCIP) (BM), 0.3 U/μl (as described in Section 14-1)

10× alkaline phosphatase buffer

0.5 M Tris, pH 9.0

10 mM $MgCl_2$

1 mM $ZnCl_2$

METHODS

Ligation Procedures for Plasmid and Insert

The procedure for ligation will vary depending on the source of insert DNA. If DNA is from a bacteriophage with identical RE ends, use method A. If RE ends are different and DNA fragment termini are asymmetrical, use method B. If the insert is cDNA from λgt10 or λgt11, use method C.

Prepare Linearized, Dephosphorylated pBR322 for Symmetric Subcloning

1. In a 1.5-ml microfuge tube, cut 5 μg of pBR322 with the desired RE, as described in Section 5-4, in a total volume of 20 μl for 60 min at 37°C.[1]
2. Add to the tube:

10× alkaline phosphatase buffer	5 μl
H_2O	24 μl
Alkaline phosphatase	1 μl

3. Briefly centrifuge all liquid to the bottom of the tube. Incubate for 30 min at 37°C.
4. Extract DNA with SS-phenol/chloroform and ethanol precipitate, as described in Section 20-1.
5. Resuspend DNA pellet in 50 μl of TE buffer. Store aliquotted samples at −20°C.

Method A: Symmetric Cloning

6. Genomic DNA fragments, cut with one RE and isolated from bacteriophage, can be visualized on a minigel (Section 9-1) using ethidium bromide. After electrophoresis, these RE fragments of interest can be isolated by electroelution (Section 9-3). The insert and prepared pBR322 plasmid can be cut with the same RE, such as *Eco*RI, and are thus ready for ligation.[2]

---

[1] It is useful to confirm that the vector DNA has been completely digested by running a small aliquot on a minigel, before proceeding to the alkaline phosphatase reaction. Incomplete digestion of the vector results in a high background of colonies without inserts.

[2] The molar ratio of insert to plasmid should be about 4 : 1.

7. Add together:

0.3 μg of the cut insert (step 6)	2 μl
0.1 μg prepared pBR322 (from step 5)[3]	1 μl
T4 DNA ligase	1 μl
10× ligase reaction buffer	2 μl
H_2O	14 μl

8. Incubate at 14°C for 4 hr to overnight.

9. Ligated insert is now ready for transformation of bacteria (Section 8-1), followed by colony screening (Section 15-3) or miniprep analysis (Section 8-4) to identify subclones with the desired insert.

Method B: Asymmetrical Cloning

10. When different REs are used to prepare insert, asymmetrical ends occur. Prepare DNA fragment by digesting with the two REs of choice. RE-digested insert is run on an agarose or acrylamide gel (Section 9-2 or 9-4) and the insert fragment is isolated by electroelution (Section 9-3).

11. Prepare 5 μg of pBR322 by cutting with compatible REs (Section 5-4). Inactivate REs by SS-phenol:chloroform extraction and ethanol precipitation, as in Section 20-1. Resuspend the purified DNA pellet in 20 μl of TE buffer.

12. Isolate desired plasmid vector fragment on an agarose or acrylamide gel (Section 9-2 or 9-4). Electroelute the appropriate fragment (Section 9-3). Resuspend this purified vector at 0.1 μg/μl in TE buffer.

13. Add together in a 1.5-ml microfuge tube:

0.4 μg of DNA insert (step 10)	2 μl
0.1 μg of plasmid DNA (step 12)	1 μl
10× ligase buffer	2 μl
T4 DNA ligase	1 μl
H_2O	14 μl

14. Incubate for 4 hr to overnight at 14°C.

15. Ligated insert is now ready for transformation of bacteria (Section 8-1), followed by colony screening (Section 15-3) or miniprep analysis (Section 8-4) to identify subclones with the desired insert.

Method C: Subcloning Inserts: "Shotgun" Cloning[4]

16. Because the inserts from a bacteriophage cDNA library may be small, they may be difficult to visualize and isolate from a gel stained with ethidium bromide. Their presence and size can usually be determined by Southern

---

[3] See note 2.

[4] See note on page 226.

blotting and hybridization to a radiolabeled probe made to the clone of interest, followed by autoradiography. The cDNA insert present in the gel may be too small a quantity for conveniently preparing fragments for subcloning. The following procedure is recommended:

17. Digest 0.5–1 μg of insert-containing λgt10 or λgt11 with *Eco*RI, in a 40 μl reaction (Section 5-4).

18. After incubation, heat the RE reaction at 70°C for 5 min. Add to 40 μl of the heat-inactivated digest:

10× ligase reaction buffer	5 μl
T4 DNA ligase	1 μl
*Eco*RI-digested, dephosphorylated pBR322 as prepared in method A (0.1 μg/μl) (step 5)	1 μl
H_2O	3 μl

19. Incubate for 4 hr to overnight at 14°C.

20. The large λgt10 or λgt11 arms will not be accepted by the plasmid for insertion, nor will pBR322 self-circularize due to the alkaline phosphatase preparation. The cDNA insert is efficiently ligated to the vector under these circumstances.

21. Take the entire volume of ligation reaction (50 μl) and proceed to Section 8-1 for transformation, followed by selection of insert-containing clones using colony hybridization with a probe, as in Section 15-3.

---

[4] A similar "shotgun" technique can be used for subcloning insert fragments from DNA in λ bacteriophage to avoid preparing the purified fragments. The bacteriophage DNA (0.5-1 μg) is digested with the RE and ligated to about 0.1 μg of an RE-cut vector. An RE should be chosen to generate fragments shorter than 10 kb for efficient cloning into pBR322. The ligation mixture is used to transform competent bacterial cells (Section 8-1). Selection of the subclones of interest is by miniprep analysis of individual transformed colonies, or by colony hybridization (Section 15-3).

SECTION **15-3.**

pBR322 Colony Hybridization

DESCRIPTION

This is a method for hybridizing a radiolabeled probe to bacterial colonies containing plasmids allowing the selection of colonies that have a plasmid with the desired insert. The colonies are partially transferred from an LB antibiotic plate to a NC filter. The colonies on the NC filter are lysed and the denatured DNA is baked onto the filter. Hybridization to the probe is identical to that described in Section 7-2. Selected colonies are then grown in selective media, and plasmid DNA is prepared for further analysis.

TIME REQUIRED

4 hr before hybridization
Overnight hybridization
2 hr on second day

SPECIAL EQUIPMENT

Vacuum oven
Plastic bag sealer
Autoradiography equipment

REAGENTS

LB ampicillin or LB tetracycline plates (Section 20-6)
NC filters
0.5 M NaOH
1.5 M NaCl with **0.5 M Tris, pH 7.4**
1.5 M NaCl with **2× SSC**
Hybridization buffer N or S, depending on probe used

METHODS

In Advance

Clone DNA of interest into pBR322 (Section 15-2). Transfect and plate colonies containing plasmids, as described in Section 8-1. Place individual colonies into a "grid" pattern on a new LB antibiotic plate by transfer with a sterile toothpick. Grow grid plates for 8–16 hr to regenerate colonies.

Prepare a nick-translated, restriction fragment probe for hybridization to the DNA of interest (Section 6-1 or 7-1). Synthetic probes may also be used.[1] Label the synthetic probe using T4 polynucleotide kinase, as described in Section 6-2.

Transfer to Nitrocellulose

1. Place a circular, dry NC filter on the grid plate containing colonies to be screened.[2,3] Put one edge down first, and drop filter to lay evenly over agar.
2. Mark filter and agar by making holes with a syringe needle around the edges of the filter in an asymmetrical pattern. Mark the bottom of the plate at the pinhole locations with a marking pen. This is helpful for later reorientation.
3. When completely wetted (about 1–2 min), remove NC filter by picking up its edge with a blunt pair of forceps at the opposite side from where the filter was placed down. Peel off filter from the plate. Some of each colony will adhere to the NC filter; a small amount will remain behind on the agar plate.
4. Reincubate the plate for 4–6 hr at 37°C to reestablish the colonies. Store plates at 4°C until positive colonies have been identified from the hybridization.

Prepare Filter

5. Place lifted NC filter, colony side up and numbered side down, on a piece of Whatman 3MM paper saturated with 0.5 M NaOH.
6. Let NC filter wet completely (5 min).
7. Remove filter and place briefly on paper towel to blot excess liquid.

---

1 Synthetic probes may be more difficult to use due to non-specific adherence to colony proteins attached to the NC filter. This background may be somewhat reduced by using Colony/Plaque Screen (NEN) filters to do the lifts and including 0.1% SDS in the hybridization buffer and washes. Note that these filters are not baked.

2 Prenumber NC filter with pencil or pen to correspond to plate number. Place filter on plate with numbered side up.

3 To distinguish nonspecific hybridization from specific binding of the probe, it is useful to include colonies in screening that are known not to contain the insert of interest (e.g., pBR322 plasmid with no cloned insert) as negative controls in the screening process.

8. Transfer filter, colony side up, to a new sheet of 3MM paper soaked in 1.5 M NaCl with 0.5 M Tris, pH 7.4. Incubate for 5 min at room temperature. Remove filter from paper and place on a clean paper towel to blot off excess buffer.
9. Transfer filter, colony side up, to a third sheet of 3MM paper soaked in 1.5 M NaCl and 2× SSC. Incubate for 5 min. Remove filter from paper. Blot excess liquid on a clean paper towel.
10. Air-dry filter on bench top for 1 hr. Bake for 2 hr at 80°C under vacuum.
11. Filters are now ready for hybridization to a radiolabeled probe. If a synthetic probe is used, go to Section 6-3. If a nick-translated probe is used, go to Section 7-2.
12. Autoradiograph hybridized NC filters, as described in Section 20-5. Realign autoradiogram with plates, using pinhole markers for orientation. Identify positive hybridizing colonies and grow colonies of interest in miniprep analysis (Section 8-4) or preparative plasmid growth (Section 8-2 or 8-3).

SECTION

15-4. Subcloning into pUC Plasmids

DESCRIPTION

An alternative to pBR322 cloning of DNA is to use pUC plasmids. These are modified pBR322 vectors with the ampicillin resistance gene and an added polylinker site, similar to that in the M13 mp vectors.[1] Also included in pUC plasmids is the 5′ end of the *lac* Z gene, which is disrupted by insertion of cloned fragments into the polylinker site, allowing a color selection of insert-bearing fragments. Insert-bearing plasmids will fail to complement the gal^- genotype of an α-complementing *E. coli* host. The plasmid transformant remains gal^-, and cannot metabolize Xgal to a blue pigment. Resulting colonies, grown on a LB/ampicillin plate supplemented with IPTG and Xgal will remain white. In contrast, noninsert-bearing plasmids complement the *E. coli* host, become gal^+, and form blue colonies on LB/ampicillin plates with IPTG and Xgal. Since a number of unique cloning sites are available in the polylinker region of pUC vectors, the pUC plasmid is often a more versatile subcloning vector than is pBR322.

The methods used for ligation and transformation using pUC are similar to those described for M13 subcloning (Sections 16-2 to 16-4), and pBR322 subcloning (Sections 15-2 and 8-1).

Cloning Inserts into pUC

DNA fragment preparation and ligation are performed as described in Section 15-2 for pBR322. Note that the available cloning sites are different, as shown in the maps in Section 4. To use the color indicator selection properties of the pUC vectors, transform and grow the plasmid in an α-complementing *E. coli* host (e.g., JM103). Plate $CaCl_2$ transformants on an LB/ampicillin plate supplemented

[1] There are many members of the pUC family. Generally, the higher designation numbers (e.g., pUC 18 versus pUC 8) have more RE sites that have been constructed into the universal cloning region. Many pairs of pUC plasmids in sequence, such as pUC 18 and 19 or 8 and 9, differ only in the left-to-right orientation of RE sites in the cloning region, similar to those described for M13 (Section 16-1).

with Xgal, IPTG and ampicillin. Follow directions for making plates as described in Section 20-6.

Growth of Plasmids

These plasmids are grown under conditions identical to those of pBR322 (Sections 8-1 to 8-4). Only ampicillin can be used for selection, because pUC has no tetracycline resistance gene. However, DNA inserts are ligated into the polylinker region.

Sequencing from DNA Cloned into pUC

Although not often used, the pUC clones can be used for partial sequencing, with procedures similar to those employed for sequencing in M13 (Section 16-10). The method used to sequence from pUC plasmids is described below.

REAGENTS

TE buffer

SS-phenol

Chloroform

Ethanol

RE

RE buffer (Section 5-4)

Universal primer (17-mer). (Collaborative Research) diluted to 250 μl with H_2O, stored at −20°C.

10× polymerase reaction buffer, stored at −20°C. See Section 16-10 for sequencing reagents.

METHODS

1. Cut 10 μg of the pUC plasmid containing the insert of interest with an RE (Section 5-4) that cuts outside of insert and primer sequences but within the polylinker region (see RE map of pUC plasmid in Section 4).
2. Perform an SS-phenol/chloroform extraction and ethanol precipitation (Section 20-1).
3. Redissolve the pellet in 20 μl of TE.
4. Combine 1 μg (2 μl) of the RE-cut plasmid with a 10-fold molar excess of the universal primer, 1.5 μl 10× polymerase reaction buffer, and H_2O to a final volume of 9.5 μl.
5. Heat contents to 95°C for 2 min. Place immediately on ice for at least 5 min to allow primer to anneal with minimal double strand formation.[2]

[2] This can also be accomplished by treatment with alkali (0.5 M NaOH). See Zagursky et al. reference.

6. Proceed to step 2 of Section 16-10 for sequencing conditions.
7. This partial sequence information can be useful in confirming that the plasmid of interest has been obtained.

REFERENCE

Ruther, U., and Muller-Hill, B., EMBO J. *2*:1791, 1982.

Zagursky, R., Baumeister, K., Lomax, N., and Berman, M., Gene Anal. Tech. *2*:89, 1985.

SECTION

16

M13 Cloning and Sequencing

SECTION **16-1.**

M13 Cloning and Sequencing: General Notes

DESCRIPTION

The next 10 sections in this manual describe an important set of techniques in molecular biology: the use of M13 vectors for the cloning and sequencing of DNA. This section provides a general description of the M13 vectors and the dideoxy sequencing technique.

A number of bacterial strains have been developed and are currently in use by molecular biologists for preparing M13 mp clones. Three of these strains are potentially useful for the methods outlined in this section; JM103 is a commonly used bacterial host for mp clones.[1] Many laboratories have used this host to successfully generate large numbers of M13 subclones useful for S_1 mapping and DNA sequencing. This strain grows well, transforms at high efficiency, and produces an acceptable yield of DNA from several milliliters of culture for most experiments. However, this strain contains two active restriction modification systems, the *Eco*K12 and *Eco*P1 systems, in contrast to its originally published genotype suggesting that it should be devoid of restriction systems. These two restriction systems may make it difficult to clone unmodified DNA that contains large numbers of the *Eco*K12 recognition site (AGACC) or the *Eco*P1 restriction site ($AACN_6GTGC$).

Two new strains have been constructed in an attempt to overcome the limitations of JM103. The *E. coli* strain JM107 can be used successfully to grow M13 mp phages. This cell line is missing in both the *Eco*K12 and *Eco*P1 restriction systems. The JM107 host grows more slowly than JM103 and yields of M13 DNA obtained in preparative growth appear to be lower than those obtained

---

[1] JM103 has a number of mutations that make it an appropriate host cell for propagation of mp-derived M13 bacteriophage vectors. The *lac* and *pro* regions of the chromosomal genome are deleted, making this host lac^- and pro^-. The proline deletion is complemented in *trans* by an F episome, which carries a proline gene and confers the male phenotype. If this episome is lost, M13 phage cannot efficiently infect a nonmale, pilus-deficient host, leading to an inefficient phage infection. To select for episome retention, JM103 stocks should be maintained on a minimal agar plate, where a pro^- phenotype (loss of the F episome) will be selected against.

with JM103. The strain JM109 is identical to JM107 but is *rec*A$^-$, lowering the probability of homologous recombination and deletions including cloned sequences. The theoretical advantages of JM107 and JM109 over JM103 should be more significant when a large insert is being subcloned, and the probability of that insert containing numerous *Eco*P1 or *Eco*K12 recognition sites is high. Any of these strains can be used for the methods described in the next few sections. Depending on the particular subcloning problem encountered, it may be advantageous to use one particular host.

Description of M13 Vectors

A number of M13 cloning vectors have been constructed and are commercially available. The M13 vectors mp8, mp9, mp10, mp11, mp18, and mp19 are circular, single-stranded DNA, filamentous, male-specific *E. coli* bacteriophage. They have been modified to include a polylinker region, with multiple unique RE cleavage sites to facilitate rapid cloning, inserted within the β-galactosidase gene (*lac Z*) in the vector. Insert-containing M13 phage are therefore gal$^-$, because the *lac Z* gene is interrupted. They form colorless plaques when growing in an appropriate α-complementing gal$^-$ host *E. coli* (e.g., JM103 or JM107) in the presence of Xgal. Nonrecombinant M13 phage used in these experiments can complement these gal$^-$ host *E. coli* and form blue plaques, allowing easy distinction from the clear plaques formed by recombinant phage.

The multiple RE sites in the polylinker region allow a variety of restriction fragments to be cloned into the M13 vectors. The various mp vectors contain different RE sites in their polylinker region; generally, more sites exist with a higher mp number in the series. In addition, the order of the RE sites in the polylinker region is directionally reversed in certain pairs of mp vectors, such as in mp8 compared to mp9 or mp10/mp11 and mp18/mp19. This pairing allows the cloning of both orientations of fragments with asymmetrical ends (e.g., *Eco*RI and *Hin*dIII fragments can be cloned in both orientations using both mp18 and mp19 vectors). This allows cloning and sequencing of one strand of a restriction fragment in mp18 and the complementary strand in mp19. Immediately adjacent to the polylinker region is a sequence that anneals specifically to a synthetic oligonucleotide primer. Thus, primer extension using large-fragment DNA polymerase I allows the cloned insert to act as a template.

In order to clone RE fragments into mp vectors, it is necessary to generate ends in the polylinker region that are compatible with ligation to the fragment ends. For example, any blunt end (an RE cleavage site that does not have either a recessed or a protruding single-stranded portion) can be ligated directly to any other blunt end. RE cleavage sites with cohesive ends (recessed or protruding) must be able to base pair with their ligation partner in such a way as to juxtapose the 3′OH group from one site with the 5′ phosphate group of the other. Any RE site can ligate to another cleavage end generated by the same RE and is therefore compatible for ligation. Some pairs of different REs also generate compatible ends;

one example is *Mbo*I; ▼GATC / CTAG▲ and *Bam*HI; G▼GATC C / C CTAG▲G.

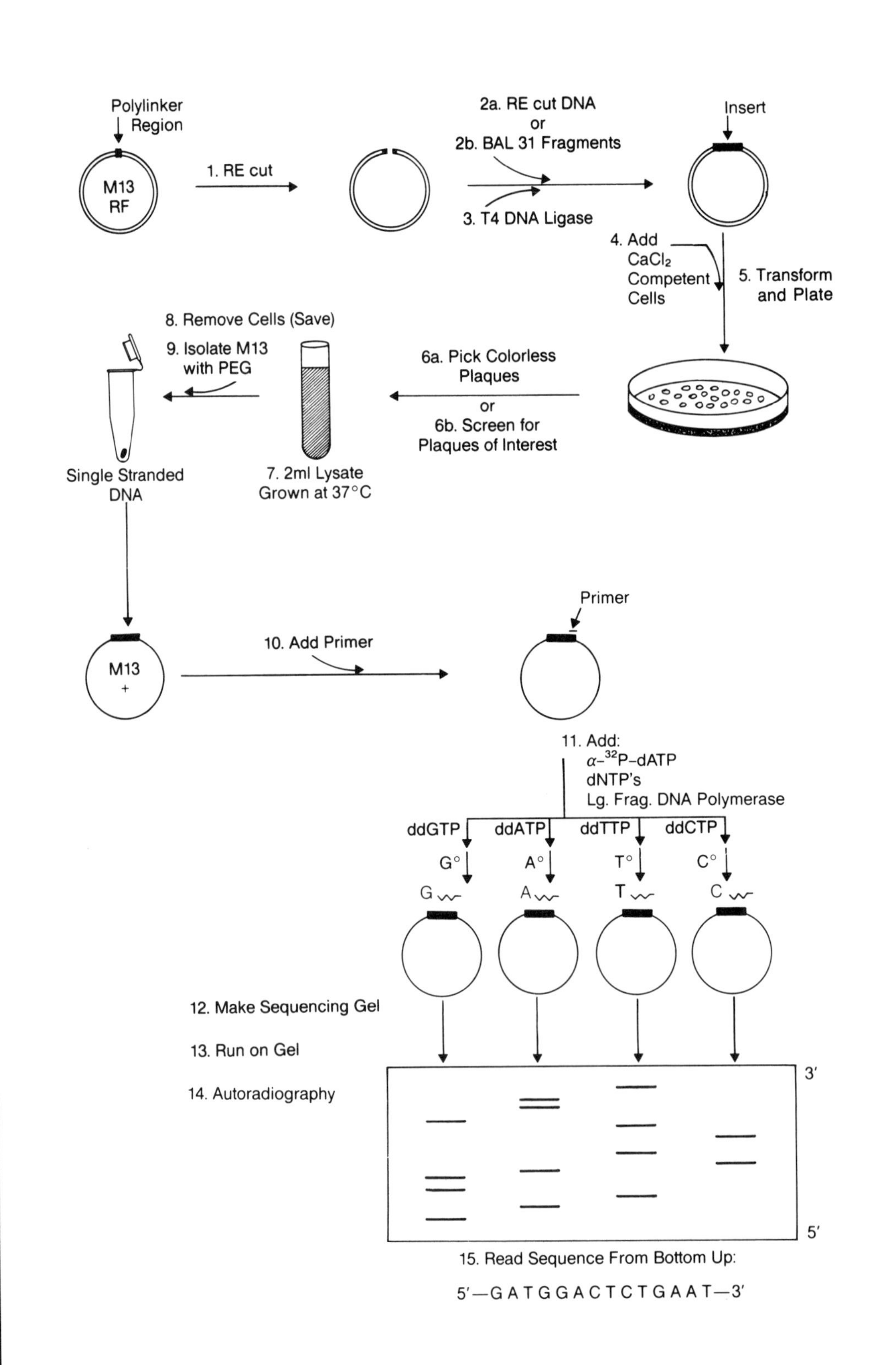
Polylinker
Region
M13
RF
1. RE cut
2a. RE cut DNA
or
2b. BAL 31 Fragments
3. T4 DNA Ligase
Insert
4. Add
CaCl2
Competent
Cells
5. Transform
and Plate
6a. Pick Colorless
Plaques
or
6b. Screen for
Plaques of Interest
7. 2ml Lysate
Grown at 37°C
8. Remove Cells (Save)
9. Isolate M13
with PEG
Single Stranded
DNA
M13
+
10. Add Primer
Primer
11. Add:
α-32P-dATP
dNTP's
Lg. Frag. DNA Polymerase
ddGTP
ddATP
ddTTP
ddCTP
G°
A°
T°
C°
12. Make Sequencing Gel
13. Run on Gel
14. Autoradiography
3′
5′
15. Read Sequence From Bottom Up:
5′—G A T G G A C T C T G A A T—3′

To prepare M13 cloning vectors it is necessary to prepare the double-stranded M13 replicative form (RF), because only double-stranded DNA can be cut with REs. Many aspects of handling, storing, propagating, and manipulating M13 vectors for use as cloning vehicles and for sequencing will be described in subsequent sections.

Sequencing of DNA Fragments

An important technique for increasing one's knowledge of gene structure is rapid, accurate DNA sequence analysis. The first method developed for rapid sequence analysis of cloned, end-labeled DNA fragments involved chemical modification and cleavage of specific nucleotides, followed by electrophoresis on high-resolution denaturing acrylamide gels. This chemical sequencing method was developed in 1977 by Maxam and Gilbert. This method is described in detail in the first two references at the end of this section and will not be reiterated in this manual.

An alternative method for rapid DNA sequencing, the dideoxy sequencing method using M13 vectors, will be presented in the sections that follow. There are three steps in the use of M13 clones for sequencing: generation of the clones, selection of the clones to be sequenced, and the primer extension sequencing procedure itself to deduce the order of bases in the extended strand. Three techniques for generating clones to sequence are described. If a detailed restriction map of the region to be sequenced is available, specific restriction fragments can be prepared and ligated directly into an M13 vector to clone the fragments of interest (Section 16-2). If a probe is available to select the region(s) to be sequenced, a heterogeneous restriction digest containing the region to be sequenced can be "shotgun" cloned, creating a small library of clones; some of them will harbor the fragment of interest. These clones may then be selected using a hybridization to the probe for detailed sequence analysis (Section 16-6). Finally, the most comprehensive and systematic method for generating clones is by successive BAL 31 exonuclease deletion. In this method, a family of overlapping clones is generated, which differs in the extent of deletion generated by BAL 31 exonuclease digestion from one end of the region to be sequenced. These successively deleted clones allow a sequence of the entire region to be determined by progressively sequencing clones that are more extensively de-

---

Figure 16.1
Schematic diagram of sequencing from M13 by the dideoxy chain termination method. DNA is cloned into the double-stranded RF form of M13, the vector with insert is used to transform cells and single-stranded M13 with insert is isolated. The single-stranded species is used as a template for a primer extension reaction with the large fragment of DNA polymerase I. The inclusion of specific ddNTPs into individual reactions causes the statistical chance of chain termination events at each possible length from the primer with each specific nucleotide at the terminus. The different length products are separated on a sequencing gel and the autoradiographed gel will represent the sequence by the position of bands (terminated by the ddNTP) in each of four possible lanes.

leted. This method can be used with little advance information about the restriction site map of the region of interest.

Two methods are presented for selecting clones to be sequenced. The first involves the use of a probe specific for the region to be sequenced to select M13 clones by hybridization (Section 16-6). Exact duplicate or sibling clones can then be identified by performing single-lane sequencing analysis on DNA prepared from the selected clones (Section 16-8). Single-lane sequencing is also useful for predicting which DNAs from a group of siblings will give the clearest sequence when a complete sequence analysis is performed.

Once the clones have been generated and selected, the primer extension is performed to determine the sequence of the insert. A synthetic oligonucleotide (17-mer) primer is annealed to the single-stranded M13 DNA genome immediately adjacent to the cloned insert. Four separate primer extension reactions employing large fragment DNA polymerase I are subsequently initiated. Each reaction contains all four deoxynucleotide triphosphates (dNTPs), but one dNTP is labeled with ^{32}P or ^{35}S to allow subsequent detection of elongated chains using autoradiography. In each of the four reactions, one dideoxynucleotide triphosphate (ddGTP, ddATP, ddTTP, or ddCTP) is added in low concentration. Thus, in each tube numerous primer extension reactions are taking place simultaneously; however, at any given point of the extension, a small percentage will incorporate ddNTP at the end of a newly extended chain. For this small portion of reactions in the tube, the nascent chain cannot be further elongated because the 3′ OH group is absent. However, other elongation reactions will continue until a ddNTP is incorporated. Thus, in the four tubes, there will be a mix of all possible lengths of the sequence, all starting at the primer region and extending up to the dideoxynucleotide in each specific tube. These fragments will be radioactive due to the labeled nucleotide inserted at multiple points along its length. When the reactions from the four tubes are run in adjacent lanes on a high-resolution sequencing gel, each fragment, from a few nucleotides in length to the maximal length completed by the primer extension, will be visualized as separate bands. The shortest fragments will run farthest down the gel.

Thus, for example, in the G tube with the dNTPs plus ddGTP, only those extended sequences ending with a G will be present, but all possible lengths of extended DNA ending in G will be in the tube. So, if C nucleotides exist in the DNA template inserted in the M13 vector at positions 15, 17, 21, and 28 past the primer (position 1), the G tube will have fragments of dideoxynucleotide terminated at positions 15, 17, 21, and 28. Moreover, if six nucleotides of the template DNA contain the sequence 5′-A T T G C G-3′, the newly synthesized primer-extended DNA fragments would carry the inverse complementary sequence that could be read 5′-C G C A A T-3′. The autoradiograph of the sequencing gel would show a C signal, a G signal, a C signal, two A signals, and a T signal, as the gel is "read" from bottom to top.

This method is rapid, accurate, and simple, requiring less labor than chemical sequencing. In addition, less radioactive material and fewer hazardous chemicals are used in the dideoxy sequencing method. Autoradiographs of the gels can be read after only 1–2 days, allowing the rapid collection of sequence data. Although only 200–400 bases are typically read from any given set of reactions, there are techniques for using this method to piece together the sequences for much larger DNA fragments.

The M13 subclones generated for sequencing purposes can also be used in other gene expression studies, such as for Northern blotting or S1 nuclease mapping. For all of the above reasons, the M13 cloning and dideoxy sequencing methods are presented in this manual as the methods of choice for most sequencing studies.

Conclusion

There are three major steps in these techniques: generation of M13 recombinant clones, selection of clones to be sequenced, and the primer extension reaction/dideoxy sequencing. All of these steps will be covered in the nine sections that follow.

REFERENCES

Maxam, A., and Gilbert, W., Proc. Natl. Acad. Sci., USA *74*:560, 1977.

Maxam, A., and Gilbert, W., Meth. Enzymol. *65*:499, 1980.

Messing, J., Meth. Enzymol. *101*:20, 1983.

M13 Cloning/Dideoxy Sequencing Instruction Manual, Bethesda Research Laboratories, Gaithersburg, Md., 1980.

Sanger, F., and Coulson, A., J. Mol. Biol. *94*:441, 1975.

Sanger, F., Nicklen, S., and Coulson, A., Proc. Natl. Acad. Sci., USA *74*:5463, 1977.

SECTION **16-2.**

Preparation of Insert for Cloning from Specific Restriction Sites

DESCRIPTION

This section describes a method for preparing DNA restriction fragments for cloning into an M13 vector. The DNA fragments may have cohesive ends compatible with ligation to any combination of sites in the M13 polylinker region. Alternatively, any blunt end may be ligated to one of two blunt end cloning sites in mp8, 9, 10, 11, 18, or 19 vectors, the *Sma*I and *Hinc*II sites. Inserts for cloning can also be further prepared, if desired, by successive BAL 31 exonuclease deletion (see Section 16-3).

TIME REQUIRED

1 day

SPECIAL EQUIPMENT

Vacuum centrifuge
Minigel apparatus
Electroelution apparatus
Scintillation counter (optional)

REAGENTS

RE for M13 insertion. This will depend on the mp vector used and the properties of the DNA to be inserted.
10× RE buffer (depends on RE used; see Section 5-4)
SS-phenol
Chloroform
TE buffer

3 M sodium acetate, pH 7.4
10× polymerase buffer
DNA polymerase I, large fragment (Klenow), 5 U/μl
Yeast tRNA, 10 mg/ml
10 mM DTT
Mix of 0.5 mM each dATP, dTTP, dCTP, and dGTP, pH 7.0
T4 DNA polymerase, 5 U/μl (BRL)
10× T4 polymerase buffer
700 mM Tris, pH 7.4
100 mM $MgCl_2$
50 mM DTT

METHODS

In Advance

Prepare plasmid or bacteriophage DNA containing the region to be sequenced. Map restriction sites in the vicinity of this region to determine appropriate fragments for M13 cloning and sequencing. Optimally, RE sites will be within 100–300 bases of the regions of interest.[1]

If DNA Fragment to Be Sequenced Has RE Sites Compatible with Cloning Sites in M13

1. Digest 5–10 μg of cloned DNA with REs to produce the desired fragment (Section 5-4). If more than one RE is used, salt conditions for each RE must be compatible. If salt conditions are not the same, use REs in sequence, starting with the one requiring the lowest salt concentration.
2. Run sample on an agarose or acrylamide gel to resolve the fragment of interest (Section 9-1).[2] Visualize samples with ethidium bromide under UV light.
3. Cut out desired band from the gel. Electroelute DNA from gel piece (Section 9-3). Add 1 μl of yeast tRNA to eluted sample.
4. Extract sample with SS-phenol/chloroform. Precipitate with ethanol (Section 20-1).
5. Dry sample by vacuum centrifugation. Resuspend fragment in TE buffer at 0.1 μg/μl. DNA restriction fragment is now ready for ligation (Section 16-1).

---

1 It is frequently difficult to determine which fragment in a complicated RE digest is desired for M13 sequencing. In this instance, the entire repertoire of fragments in the digest can be ligated into a prepared vector for shotgun cloning. This generates a library of clones in M13 that contains all compatible fragments in the digest. The clones desired for sequencing may be selected by hybridization to a radiolabeled probe that defines the region of interest (Section 16-6).

2 Use agarose gels for larger fragments (over 1,000 bp) and acrylamide gels for smaller fragments. Small fragments are more easily cloned in M13 vectors.

If DNA Ends Are Not Compatible with M13 Cloning Sites

Conversion of a noncompatible end to a blunt end can be done with large-fragment DNA polymerase I (for 5′ protruding end) or T4 DNA polymerase (for 3′ protruding end).[3] Note that either 5′ or 3′ protruding ends can result, depending on the RE chosen. To generate blunt ends, use steps 3 to 5 for a 5′ protruding end or steps 6 to 8 for a 3′ protruding end.

1. Digest 20 μg of cloned DNA to completion with an RE whose end will be converted to a blunt form. Remove RE by SS-phenol/chloroform extraction and ethanol precipitation (Section 20-1).
2. Resuspend purified DNA digest in 20 μl of TE buffer.

For 5′ Protruding Ends

3. Mix together:

DNA sample	20 μl
10× polymerase buffer	3 μl
10 mM DTT	3 μl
dNTP mix (0.5 mM each)	3 μl
DNA polymerase I (large fragment)	1 μl

4. Incubate for 15 min at 37°C.
5. Inactivate enzyme by heating to 70°C for 10 min. Proceed to step 9.

For 3′ Protruding Ends

6. Mix together:

DNA sample	20 μl
10× T4 polymerase buffer	3 μl
dNTP mix (0.5 mM each)	3 μl
H_2O	3 μl
T4 polymerase	1 μl

7. Incubate for 15 min at 37°C.
8. Inactivate enzyme by heating to 70°C for 10 min.
9. Purify DNA by SS-phenol extraction and ethanol precipitation (Section 20-1). Resuspend DNA pellet in 20 μl of TE buffer.

---

[3] Large-fragment DNA polymerase I (Klenow) will efficiently convert any 5′ protruding RE site to a blunt end. This enzyme is less efficient in converting 3′ protruding ends to blunt ends. T4 DNA polymerase is a better choice for blunting 3′ protruding ends.

10. Noncompatible 5′ or 3′ protruding end is now blunt, and compatible with ligation into M13 mp vectors containing a *Sma*I or *Hinc*II RE site in their polylinker region.
11. If necessary, the DNA from step 9 can be digested with an additional RE to generate an asymmetrical fragment for cloning into the mp vector (see Section 5-4). After RE digestion, purify DNA by extraction and precipitation, as described in Section 20-1, and resuspend pellet in 20 μl of TE.
12. The desired fragment may now be purified by resolution on an agarose or acrylamide gel (Section 9-2 or 9-4), electroeluted (Section 9-3), and the eluted fragment resuspended at 0.1 μg/μl in TE buffer. Alternatively, the entire digest can be ligated into an appropriate M13 vector ("shotgun cloning"), generating a library of clones (Section 16-4). Hybridization of the clones to a radiolabeled probe for the DNA of interest can be used to select the M13 clones to be sequenced (Section 16-6).

SECTION **16-3.**

Preparation of Insert for M13 Cloning by Successive BAL 31 Exonuclease Deletion

DESCRIPTION

This method is used to prepare a series of systematically deleted fragments covering the region to be sequenced in M13. These fragments, to be cloned into the M13 polylinker region, are cut at defined distances away from a single RE site using the exonuclease BAL 31. M13 cloning using this strategy allows one to "walk" across the sequence of a region by analyzing overlapping, systematically deleted M13 clones. This method involves six steps: 1) a DNA fragment to be sequenced is subcloned into a pBR plasmid vector, 2) preparative growth of the plasmid/insert, 3) cleavage of the plasmid by digestion with one RE, 4) preparation of successively shorter samples by BAL 31 exonuclease digestion, 5) polishing of the BAL 31–digested ends to form blunt ends, and 6) removal of the series of progressively deleted inserts from plasmid DNA by digestion with a second RE. After the insert fragments are prepared by gel electrophoresis and electroelution, this family of inserts is ligated into M13 mp vectors, and the overlapping sequences that are derived allow determination of the entire sequence of the DNA of interest.

TIME REQUIRED

Several days

REAGENTS

RE (see below for selection)
10× RE buffer (Section 5-4)
SS-phenol
Chloroform
3 M sodium acetate, pH 7.4
10× polymerase buffer

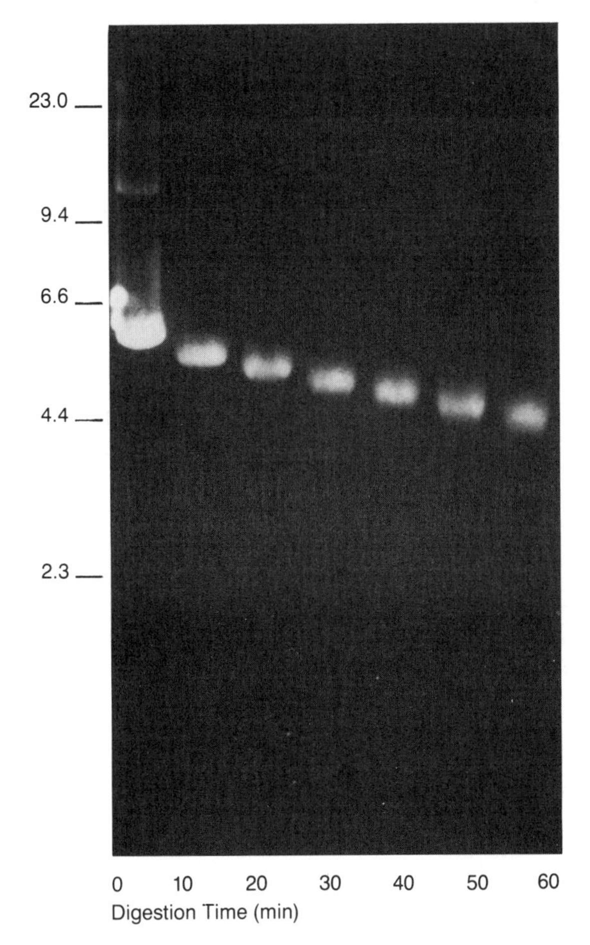

Figure 16.2
Photograph of ethidium bromide staining of DNA in an agarose gel following successive BAL 31 digestion. Note the progressive decrease in size of the bands with increased time of exposure to the enzyme (left to right).

DNA polymerase I, large fragment, (Klenow), 5 U/μl
dNTP mix (0.5 mM each dGTP, dATP, dTTP, dCTP) pH 7.0
BAL 31 enzyme, 3 U/μl (NEBL)
5× BAL 31 buffer
  3 M NaCl
  60 mM $CaCl_2$
  60 mM $MgCl_2$
  100 mM Tris, pH 8.0
  5 mM EDTA

Yeast tRNA, 10 mg/ml
0.5 M EGTA, pH 8.0
10 mM DTT
TE buffer

METHODS

In Advance

Subclone a DNA fragment with asymmetric termini to be sequenced, up to 2 kb in length, into a plasmid vector, using RE sites that are compatible with ligation in the M13 polylinker region (e.g., *Eco*RI-*Bam*HI, *Bam*HI-*Hin*dIII). See Sections 15-1 to 15-3 for plasmid subcloning. Grow the plasmid preparatively (Section 8-2 or 8-3).

Procedure

1. Digest 100 μg of the plasmid subclone grown for M13 subcloning with one of the two REs used in its asymmetrical cloning. Perform digest in 250 μl volume (Section 5-4).
2. Check an aliquot of the sample on a minigel to verify complete digestion (Section 9-1).
3. Transfer 23 μl of digest to a microfuge tube and place on ice. Freeze remaining 227 μl at −20°C.
4. To 23 μl of digest add:

   6 μl of 5× BAL 31 buffer
   1 μl of BAL 31 enzyme

5. Incubate for up to 1 hr at 30°C, using the following regimen: Remove 5-μl aliquots every 10 min to a new tube. Stop BAL 31 reaction by *immediately* adding to the aliquot 1 μl of 0.5 M EGTA, pH 8.0, and freezing on dry ice.
6. When all samples are frozen, add 1 μl of yeast tRNA carrier and 20 μl of H_2O to each sample.
7. Purify DNA in each tube by SS-phenol/chloroform extraction and ethanol precipitation (Section 20-1).
8. Resuspend pellet in 10 μl of TE buffer. Compare the mean size of digested fragments at each time point by agarose gel electrophoresis, against *Hin*dIII-cut λ DNA size markers (Section 9-2).
9. From the determined rate and extent of deletion, judge the time needed to digest the subcloned insert appropriately under the conditions in step 4.
10. From the sample in step 3 remove a 25-μl aliquot (time 0). With the remaining sample set up a preparative BAL 31 digestion reaction:

Frozen DNA digest (step 3)	200 μl
5× BAL 31 buffer	60 μl

BAL 31 enzyme (3 U/μl)	10 μl
H_2O	30 μl

11. Incubate at 30°C. Remove ten 30-μl aliquots at appropriate times to generate a range of deletions covering the region of interest. Immediately terminate the reaction in each aliquot by adding 3 μl of 0.5 M EGTA, pH 8.0, and freezing on dry ice.
12. Pool the time points and purify DNA by SS-phenol/chloroform extraction followed by ethanol precipitation (Section 20-1).[1]
13. Resuspend pellet in 50 μl of TE buffer.
14. Polish the deleted fragments by filling in any nonblunt ends with large-fragment DNA polymerase I. Add together:

DNA from step 13	50 μl
10× polymerase buffer	10 μl
dNTP mix (0.5 mM each)	10 μl
DNA polymerase I (large fragment)	5 μl
H_2O	15 μl
10 mM DTT	10 μl

15. Incubate for 15 min at 37°C. Heat inactivate at 70°C for 10 min.
16. Adjust buffer to conditions favorable for digestion with the RE whose site was not subjected to the BAL 31 exonuclease digestion in the plasmid subclone used (see "In Advance"). Usually this involves diluting DNA into an appropriate buffer (e.g., 400 μl volume), as described in Section 5-4.
17. Digest DNA with the second RE to completion.
18. Purify DNA by extraction with SS-phenol/chloroform and ethanol precipitation (Section 20-1).
19. The insert of interest now exists as a population of deleted molecules of varying lengths, depending on the extent of digestion. These fragments share a common end, generated by the second RE digestion, and all have a blunt end.
20. The insert is now size fractionated on an agarose gel (Section 9-2) or an acrylamide gel (Section 9-4). Gel pieces containing fragments whose size will allow overlapping sequencing (progressively shorter by about 200 bp) are cut out and electroeluted (Section 9-3).[2] Estimate size by comparison with size markers run on the same gel. Electroelute enough gel pieces to give a series of overlapping clones whose average size will differ by 200–250 bp in length, and together will cover the entire DNA insert.

---

[1] The collected time points can be treated individually to assure representative clone selection covering the entire sequence. Follow same procedure using the 10 or 11 samples. The size-selected pieces can also be obtained in step 20.

[2] tRNA carrier (1 μl of 10 mg/ml stock) may be needed during electroelution to reduce losses of sample.

21. After electroelution, resuspend insert samples at approximately 0.1 μg/μl in TE buffer.

22. The inserts are now ready for ligation into an appropriate M13 mp vector, asymmetrically digested such that the blunt (BAL 31–deleted) end of the cloned insert is closest to the universal primer site (Section 16-4). Blunt sites compatible with BAL 31–deleted blunt ends are *Sma*I and *Hin*cII. The second site for vector ligation must be compatible with the second RE site common to all insert fragments.[3,4]

---

[3] An example of insert generation and vector choice is as follows:

Plasmid subclone: 1.5-kb *Eco*RI-*Hin*dIII insert.

First cleavage: *Eco*RI and then BAL 31 deletion and conversion to blunt end.

Second cleavage: *Hin*dIII then electroelution.

Final product: Insert with one blunt end and one *Hin*dIII-cut end.

M13 vector: mp 9, mp 11, or mp 19 cut with *Sma*I and *Hin*dIII.

[4] After cloning, it is usually helpful to screen potential clones of interest with a probe specific for the region to be sequenced (Sections 7-1 and 7-2). DNA can then be prepared from hybridizing clones for sequence analysis.

SECTION **16-4.**

M13 Vector Preparation and Ligation of Insert into Vector

DESCRIPTION

This method describes the preparation of double-stranded, replicative form (RF) M13 bacteriophage and the digestion of the double-stranded phage DNA into a linear vector with ends compatible for ligation to the prepared insert. Following vector preparation, ligation of the prepared insert is described.

TIME REQUIRED

4 days to prepare RF form of M13 mp vector

1 day to prepare vector for ligation and to ligate insert

REAGENTS

DNA insert, prepared for cloning (see Sections 16-2 and 16-3)

M13 RF DNA for mp cloning vectors 0.1 μg of vector (e.g., purchased from BRL or previously prepared)

LB medium

JM103 or JM109 colony growing on minimal agar plate[1]

---

1 JM103 cells are recommended here, but other cells can be chosen (see Section 16-1). JM103 has a number of mutations that make it an appropriate host for propagation of mp-derived M13 bacteriophage (see Section 16-1). For optimal yields of DNA and RF in preparative growth steps, the JM103 cells can be grown in richer medium, such as LB or 2× YT. It is dangerous, however, to allow the JM103 cells to grow for many generations in a rich medium, because F episome loss may occur. If this happens, reinfection of JM103 with M13 phage will be prevented. Thus, we recommend that JM103 stock plates be maintained on M9 minimal agar plates and that these plates be restreaked every 2–4 weeks. When working with these cells for preparative growth of M13 DNA or transfection, JM103 may be grown for a short period in a richer medium, but prolonged stationary growth in this medium should be avoided.

Chloramphenicol, 30 mg/ml, in ethanol

2× YT medium

Bacto-tryptone	16 g
Bacto yeast extract	10 g
NaCl	10 g

H_2O to 1 liter, pH to 7.4

Sterilize by autoclaving.

REs for vector preparation; depends on insert prepared (see Section 16-2 or 16-3)

10× RE buffer(s)

0.2 M EDTA, pH 8.0

SS-phenol

Chloroform

3 M sodium acetate, pH 7.4

Ethanol

10× ligase buffer

T4 DNA ligase

0.1 M DTT

METHODS

In Advance

A strategy for preparing the M13 insert should also include the preparation of the M13 vector. Once the strategy for cloning is determined, the mp vector with RE sites compatible with the prepared insert ends must be selected and prepared. The chosen mp vector should have an RE site in the polylinker region such that the region to be sequenced will be as close as possible to the universal priming site. For a list of RE sites on mp vectors, see Section 4.

Note: The mp vector RF form can be purchased in quantity. If purchased RF will be used in M13 ligation, proceed to step 8.

On the day before inoculation, transfect 0.1 μg of the selected M13 RF into competent JM103 cells, producing a plate of blue plaques, as described in Section 16-5.

To Prepare M13 RF

1. Sterilize two 500-ml Erlenmeyer flasks and one 4-liter flask containing 900 ml of 2× YT medium by autoclaving.
2. Put 100 ml of LB medium in one sterile 500-ml flask. Inoculate with one colony of JM103 cells from an M9 minimal agar plate.[2] Grow, with shaking, at 37°C to an O.D. at 600 nm of 1.0. Store these cells at 4°C.

---

[2] JM103 bacteria stocks are maintained and stored on minimal plates to prevent loss of the F episome.

3. Place 100 ml of LB medium in the second sterile 500-ml flask. Inoculate this medium with one blue plaque from the mp vector plate by transferring a blue plaque in an agar plug in the tip of a sterile Pasteur pipette. Grow, with shaking, at 37°C overnight to make a phage stock.

Day 2

4. Add 100 ml of LB containing JM103 from step 2 to the 900 ml of 2× YT medium in a 4-liter flask. Grow, with shaking, at 37°C to an O.D. at 600 nm of 1.5.

5. Add the 100 ml of phage stock from step 3 to these cells to begin infection. Infect by shaking at 37°C for 15 min.

6. Add 0.5 ml of chloramphenicol solution to the infected JM103 cells. Shake at 37°C for 1–2 hr to allow accumulation of RF DNA.

7. Prepare closed circular M13 RF DNA (approximately 7-kb circles) from JM103 host, using the triton lysis method described in Section 8-2, beginning with step 9 for harvesting cells.[3,4]

Preparation of M13 Vector

8. Digest 2–5 μg of the appropriate M13 RF with REs to generate a cloning site in the polylinker region with ends compatible to the insert(s) prepared in Section 16-2 or 16-3. If the fragment to be cloned is asymmetrical, two RE sites will be needed. (See Section 5-4 for RE digestion.)

9. Incubate the M13 RF with 2–3 U of the appropriate RE per microgram of DNA for 1 hr in a 20-μl volume (with appropriate RE buffer, etc.).[5]

10. Add 1 μl of yeast tRNA solution.

11. Extract with SS-phenol/chloroform and ethanol precipitate, as described in Section 20-1.

12. Resuspend vector at 0.1 μg per microliter of TE buffer.

13. Run 0.1 μg of prepared vector on a minigel (Section 9-1) to verify complete digestion (i.e., no circular RF DNA remaining). Run 0.1 μg of undigested

---

[3] The yield of M13 RF DNA is typically 200–500 μg/liter. When purified as described in Section 8-2, the upper chromosomal, ethidium bromide band on the first CsCl vertical gradient is much more pronounced with the M13 RF preparations than when plasmid DNA is prepared.

[4] It is worth noting that recombinant M13 clones can also be grown for RF preparation. Double-stranded DNA probes can be prepared from this growth and used for nick translation, or useful restriction fragments can be made and further subcloned into another cloning vector.

[5] If two REs are needed to make the desired vector, it is convenient to digest with both REs at the same time, as long as they are both active under the same buffer conditions (Section 5-4). If one enzyme requires a lower-salt buffer, digest first with that RE for 1 hr, add salt to meet the requirements of the second RE, and then perform the second RE reaction.

DNA for comparison. Linear digestion of the circular RF decreases the mobility of the DNA on agarose gels.

Ligation of Insert into Prepared Vector

14. A typical 20-μl ligation reaction for generating M13 clones is assembled as follows:[6]

M13 vector (step 12)[7]	1 μl
Compatible insert[8]	0.5–3 μl
10× ligase buffer	2 μl
T4 DNA ligase[9]	1 μl
H_2O	To make 20 μl

15. Incubate at 14°C for 4 hr to overnight.

16. Ligations may be frozen at −20°C for storage until needed for transformation of competent JM103 cells.

---

6 If only one RE was used it is frequently useful to perform a control ligation, with no added insert, in parallel with the ligation of the insert. In this control, only plaques from M13 recircularization should result, and they should be blue. If there is a significant background of clear plaques on the control plate, the REs used to cut the M13 RF are damaging the cloning site and inactivating the *lac Z* gene, or there is foreign DNA in the M13 vector preparation.

7 If the vector used is symmetrical and blunt (e.g., *Sma*I cut), use two- to threefold more vector and DNA to compensate for the inefficiency of T4 ligase in blunt ligation.

8 The amount of compatible insert used depends on its size. In general, it is useful to have approximately three to four times the molar amount of the insert compared to the vector for optimal formation of clones.

9 T4 ligase requires magnesium for optimal activity. Be careful not to add excessive EDTA in this reaction in the vector and insert solutions.

SECTION

16-5. Transformation of M13 into JM103 *E. coli* Host

DESCRIPTION

This section describes a method for transforming *E. coli* cells, such as JM103, with a ligation mix containing recombinant M13 vector DNA and an insert. Recombinant M13-infected centers, or plaques, are produced, which can be used to grow the M13 recombinant phage preparatively. DNA is then prepared from these phage for subsequent analysis.

TIME REQUIRED

1 day

SPECIAL EQUIPMENT

Heating block apparatus for plating bacteriophage

REAGENTS

LB medium

50 mM $CaCl_2$, sterilized by autoclaving

LB top agar

Xgal solution, 10% in DMF

IPTG, 100 mM in sterile H_2O

LB agar and minimal agar plates (see Section 20-6)

METHODS

In Advance

Prepare LB medium. Streak out JM103 *E. coli* cells on a minimal agar plate by spreading a small volume of stock cells with a bent glass rod or wire loop over surface of plate. Grow JM103 colonies overnight at 37°C with the plate agar-side up. Prepare ligated M13 clones, as described in Section 16-4.

Preparation of Competent JM103 Cells

1. With flame-sterilized loop, scrape off several colonies of JM103 bacterial cells. Add cells to 50 ml of LB medium in a sterile culture flask.
2. Incubate cells at 37°C with shaking until the O.D. at 600 nm is approximately 0.3. This takes approximately 2–4 hr. Remove 2 ml of cells and save in a sterile tube at 4°C.
3. Spin down remaining cells for 5 min at 3,000 × *g*.
4. Discard medium. Resuspend cell pellet gently in 2–4 ml of 50 mM $CaCl_2$. Bring volume up to 20 ml with this $CaCl_2$ solution.
5. Incubate for 20 min on ice.
6. Centrifuge cells for 5 min at 3,000 × *g* at 4°C. Discard supernatant. Resuspend cell pellet in 2.5 ml of ice cold 50 mM $CaCl_2$.
7. Store cells at 4°C for up to 48 hr until ready for transformation. These cells are now competent and can be stored on ice for at least 2 days.

Transformation of Competent Cells

8. To achieve an optimal density of single plaques for isolation of sufficient numbers of clones, two aliquots of each M13 ligation (from Section 16-4) are plated: 2 μl and 8 μl. The remaining 10 μl of each ligation is stored frozen at −20°C in case more plaques are needed. Add the 2-μl and 8-μl aliquots to sterile 12 × 75 mm glass tubes containing 150 μl of competent JM103 cells from step 7. Incubate for 40 min on ice.
9. Melt LB top agar using an autoclave or microwave oven. Allow to cool to 48°C. After cooling, add 100 μl of 10% Xgal (in DMF) and 100 μl of 0.1 M IPTG for each 30 ml of top agar (enough for 10 platings). Maintain this mixture at 48°C until needed.
10. Heat-shock JM103/M13 mixture from step 8 for 5 min at 37°C or for 2 min at 42°C.
11. Add 2 drops of the exponentially growing JM103 cells, saved in step 2, to each transformation tube to be plated. Transfer tubes to a heating block set at 48°C.
12. Add 3 ml of the LB top agar/IPTG/Xgal mix to each tube on the heating block.

13. Immediately pour each tube onto a 90-mm LB agar plate and swirl to create a level surface of top agar.
14. Wait 15 min for top agar to harden. Cover and invert plates.
15. Incubate plates overnight at 37°C to allow plaque formation and color indicator reaction.[1]
16. Plaques are now ready for screening prior to sequencing (see Section 16-6).

---

[1] Plaques typically start to become apparent after about 6 hr of incubation. The color change due to metabolized Xgal is not apparent until 8–10 hr of incubation. Pick plaques within the next 24 hr.

SECTION **16-6.**

Screening M13 Clones with a Radiolabeled Probe to Select Inserts for Sequencing

DESCRIPTION

This method employs a specific nick-translated probe complementary to the DNA of interest to select for a homologous DNA insert in M13 vectors.[1] This procedure is particularly convenient when the insert population used in ligation and transformation (in the previous two sections) contains only a small portion of inserts relevant for sequence analysis. Screening is a rapid and efficient technique to identify clones of interest prior to sequencing.

TIME REQUIRED

1 day

SPECIAL EQUIPMENT

96-well microtiter dishes (e.g., Costar)

96-well microtiter block transfer device (optional)

Sterile toothpicks

REAGENTS

LB medium

0.2 M NaOH with **1.5 M NaCl**

2× SSC buffer with **0.4 M Tris, pH 7.4**

---

[1] Synthetic probes can also be used for selection of clones of interest. Due to their short size, a more limited selection may occur that does not completely represent the cDNA clone if BAL 31 was used. Use hybridization buffer-S and different conditions, as described in Section 6-3 for synthetic probes.

2× SSC buffer
Hybridization buffer N[2]

METHODS

In Advance

Transform JM103 cells with M13 ligations and plate, as described in the previous section. Prepare nick-translated probe to DNA of interest to be sequenced, as described in Section 7-1.

Screening Procedure

1. Fill a 96-well microtiter plate with 150 μl of sterile LB medium in each well.
2. Transfer each well-defined, colorless plaque from a transformation plate to a microtiter plate well by touching a toothpick to the plaque and then into the well. Transfer all colorless plaques that need to be screened.[3]
3. Incubate inoculated microtiter plates at 37°C for 4–6 hr. These plates may then be stored for up to 2 weeks before screening or use for preparative growth of DNA for sequencing.
4. Using a 96-well microtiter transfer device, if available, transfer 10 μl of the infected medium from each well to an NC filter. If a block transfer device is not available, transfer spots to filter using a pipettor and sterile pipette tips. Mark the filter to allow orientation of the pattern of supernatant spots.
5. Air-dry filter for 15 min at room temperature or for 5 min at 70°C.
6. Denature and neutralize DNA on filter by dipping filter sequentially for 30 sec each in 100 ml of the following solutions:

   0.2 M NaOH with 1.5 M NaCl
   2× SSC with 0.4 M Tris, pH 7.4
   2× SSC

7. Dry NC filter by blotting with Whatman 3MM paper.
8. Bake filter in vacuum oven for 2 hr at 80°C.
9. Hybridize filter to the specific probe, as described in Section 7-2 for nick-translated probes.
10. Autoradiograph hybridized NC filter as described in Section 20-5.
11. Dark spots on autoradiograph will correspond to positive clones. The corresponding microtiter wells contain clones to be used in Section 16-7. Store microtiter plate covered at 4°C before use.

---

[2] See note 1.

[3] The number of plaques to be screened depends on the likelihood of obtaining the desired cloned insert. If the insert population available for ligation into the vector was heterogeneous, many clones will have to be tested. If the insert was isolated and purified before ligation, few clones will need to be tested.

SECTION **16-7.**

Preparation of Single-Stranded M13 DNA for Sequencing

DESCRIPTION

This method describes a protocol to grow preparatively and isolate cloned M13 DNA after transformation. The single-stranded template DNA prepared is ready for sequencing or other subsequent use (i.e., for synthesizing probes using the M13 template).

TIME REQUIRED

Day 1—1 hr
Day 2—3 hr

REAGANTS

2× YT medium

Bacto-tryptone	16 g
Bacto-yeast extract	10 g
NaCl	10 g
H_2O	to make 1 liter

pH to 7.4; sterilize by autoclaving

20% PEG/2.5 M NaCl solution

PEG-8000	20 g
NaCl	14.5 g

Add H_2O to make 100 ml.

TES buffer

20 mM Tris, pH 7.4

10 mM NaCl

0.1 mM EDTA (Na_2)

SS-phenol

Chloroform

3 M sodium acetate, pH 7.4

Ethanol

METHODS

In Advance

Prepare M13 transformants by $CaCl_2$ transformation of an appropriate ligation mix into JM103 cells (Section 16-5). Transfer colorless plaques to sterile LB medium in individual wells of a microtiter dish and screen with probe, if necessary (Section 16-6), to select clones appropriate for preparative growth and sequencing.

Inoculate 10 ml of LB medium with JM103 cells from a minimal agar plate stock. Grow cells to mid-log phase (O.D. at 600 nm of about 0.3).

Procedure

1. For each clone, add 2 ml of 2× YT medium in a sterile tube.[1] Add 20 μl of mid-log phase JM103 cells to each tube. Add 20 μl of infected LB medium from microtiter well containing a clone selected for growth and sequencing.
2. Grow infected cells for 6–16 hr at 37°C with good aeration.
3. Decant 1.25 ml of cells from step 2 into a 1.5-ml microfuge tube. Spin in microcentrifuge for 5 min at room temperature. Transfer 1 ml of the supernatant to a new tube, being careful not to disrupt bacterial cell pellet.
4. To the 1 ml of supernatant in the new tube, add 250 μl of PEG/NaCl solution. Mix by inverting tube and vortexing. Incubate for 10 min at room temperature.
5. Spin in microcentrifuge for 10 min. Decant supernatant. Wipe off inside wall of tube with a cotton-tip applicator or paper wipe to remove any residual PEG and 2× YT medium. Carefully swab last drop over pellet with a paper wipe without disturbing the PEG bacteriophage pellet.[2]
6. Add 100 μl of TES buffer. Resuspend pellet by vortexing.
7. Extract bacteriophage with 50 μl of SS-phenol. Mix by vortexing. Let stand for 5 min and mix again by vortexing. Spin sample in microcentrifuge to separate phases for 2 min.

---

[1] LB medium may be substituted for the 2× YT medium.

[2] The pellet should appear as a small spot at the bottom of the tube. If no pellet is visible at this step, the yield of bacteriophage DNA will probably be less than optimal.

8. Transfer upper aqueous phase, approximately 90 μl, to a new tube. Add 90 μl of chloroform. Mix by vortexing. Spin in microcentrifuge for 2 min.
9. Transfer upper aqueous phase (about 80 μl) to a new tube. Add 8 μl of 3 M sodium acetate, pH 7.4.
10. Add 300 μl of ethanol. Mix by vortexing. Freeze on dry ice for 10 min. Spin in microcentrifuge for 5 min at 4°C.
11. Decant ethanol. Add 200 μl of 95% ethanol. Mix by vortexing. Spin in microcentrifuge for 5 min.
12. Decant ethanol. Dry pellet in vacuum centrifuge or vacuum evaporator.
13. Resuspend DNA in 50 μl of TES buffer.
14. Check for recovery of DNA by running 2.5 μl of the sample on a minigel or agarose gel (Section 9-1 or 9-2). The phage DNA should be clearly visible with ethidium bromide staining. Large inserts (over 1 kb in length) will retard the electrophoretic mobility of the phage DNA, as compared with phage DNA without insert.[3]
15. Store samples at −20°C. DNA is very stable. This is the DNA used for primer extension in single-lane screening, sequencing, and S_1 nuclease assay. The yield will be 0.2–2 μg of DNA.

REFERENCE

M13 Cloning/Dideoxy Sequencing: Instruction Manual. Bethesda Research Laboratories, Gaithersburg, Md., 1980.

---

[3] Mobility of M13 DNA is not comparable to double-stranded DNA standards, such as *Hin*dIII-cut λ.

SECTION

16-8. Single-Lane Screen Analysis of M13 Clones

DESCRIPTION

This is a rapid method for comparing the inserts and their orientation in many M13 DNA preparations. This procedure allows an easy decision to be made about which clones should be sequenced entirely, and allows elimination of identical sibling clones that would yield no new information. By this method, a sequencing reaction with a single dideoxy nucleotide, ddGTP, is run for each M13 DNA in a manner similar to the full sequencing described in Section 16-10. Sibling clones show an identical pattern in this assay; opposite orientations of the same symmetrically cloned insert give two different patterns.

TIME REQUIRED

1 day

SPECIAL EQUIPMENT

Sequencing gel apparatus, with plates and a high-voltage power supply (e.g., 2,000 V)

Gel dryer

REAGENTS

α-^{32}P-dATP (400 Ci/mmol) or α-^{35}S-dATP (1,000 Ci/mmol) (NEN or Amersham)

M13 universal primer, 17-mer (e.g., Collaborative Research, "1,000 reaction," diluted in 250 μl with H_2O, stored in aliquots at −20°C)

DNA polymerase I, large fragment, Klenow, 5 U/μl (BRL), stored at −20°C

10× polymerase buffer

0.1 M DTT, store at −20°C

dGTP, 10 mM, pH 7.0

dATP, 10 mM, pH 7.0
dTTP, 10 mM, pH 7.0
dCTP, 10 mM, pH 7.0
Store each of the dNTPs in 50-μl aliquots at −70°C
ddGTP, 10 mM, pH 7.0; store in 5-μl aliquots at −70°C.
Sequencing gel loading solution
950 μl deionized formamide
50 μl *20× TBE buffer*
0.1 mg bromophenol blue
0.1 mg xylene cyanol
Store at −20°C in small aliquots.

METHODS

In Advance

Make an 8% polyacrylamide 8 M urea sequencing gel, 40 cm in length, as described in Section 16-9.

Procedure

1. Dilute dNTP stocks to 2 mM with H_2O.
2. Dilute ddGTP to 0.5 mM with H_2O.
3. From 2-mM dilutions of dNTP stocks, make a G° solution:

   1 μl dGTP
   20 μl dTTP
   20 μl dCTP

   Store at −20°C for up to 4 weeks.

   Also make a "chase" solution from 2-mM stocks from step 1:

   25 μl dGTP
   25 μl dATP
   25 μl dTTP
   25 μl dCTP

   Store at −20°C for up to 4 weeks.

4. Make a single-lane sequencing primer annealing reaction for each M13 DNA in a 1.5-ml microfuge tube (multiply these values by the number of tubes needed):

   0.2 μl primer
   0.5 μl 10× polymerase buffer
   1.3 μl H_2O

5. In individual 1.5-ml microfuge tubes, add 1 μl of each M13 DNA sample and 2 μl of the mix from step 4.
6. Heat to 55°C and allow to cool slowly over 1–2 hr to room temperature (25°C). This allows annealing of primer to M13 clone. After annealing, collect the 3-μl mix by centrifugation for 30 sec.
7. Make a polymerization mixture as follows (the quantities below are for 20 single-lane analyses):

   15 μl α-^{32}P-dATP or ^{35}S-dATP[1]
   20 μl G° mix
   20 μl ddGTP (0.5 mM dilution)
   1 μl 0.1 M DTT
   1 μl 10× polymerase reaction buffer
   5 μl DNA polymerase I, large fragment

8. Add 2.8 μl of the polymerization mix from step 7 to the 3 μl of each annealing mix from step 5. Incubate tubes for 10 min at room temperature.
9. Add 1 μl of the chase solution to each tube. Incubate for an additional 10 min at room temperature.
10. Add 4 μl of sequencing gel loading solution to each tube.
11. Heat tubes in a boiling water bath for 1 min.
12. Load 2 μl of each sample to individual wells on the sequencing gel.
13. Run sequencing gel at 50 W until xylene cyanol dye in sample is midway down the gel. This usually requires about 2 hr at 1,500–2,000 V. Xylene cyanol runs with species of approximately 80 bp.[2]
14. Transfer gel to a sheet of Whatman 3MM paper. A detailed method for performing this transfer is included in steps 13 to 16 of Section 16-10. Cover gel with plastic wrap and dry gel using a large gel dryer for 1–2 hr at 80°C under vacuum.[3]
15. Select clones for full sequence analysis, based on single lane pattern of G termination. Choose the best of any sibling clones for full analysis.[4]

REFERENCES

M13 Cloning/Dideoxy Sequencing: Instruction Manual. Bethesda Research Laboratories, Gaithersburg, Md., 1980.

Sanger, F., Nicklen, S., and Coulson, A. R., Proc. Natl. Acad. Sci., USA *74*:5463, 1977.

---

[1] Handle radioactive material with care. Wear plastic gloves throughout the procedure.

[2] The lower electrophoresis tank will become radioactive. Handle appropriately.

[3] For ^{35}S labeling, see Section 16-10, note 5. The plastic wrap should be removed from the gel after drying and before autoradiography. This step is not necessary for ^{32}P labeling. Saran wrap is recommended.

[4] The insert similarities will be apparent from the G bases alone. Symmetrical inserts with two different orientations will appear as two sets of sibling clones.

SECTION

16-9. Preparation of Polyacrylamide Sequencing Gel

DESCRIPTION

A 0.4-mm-thick, 8 M urea, polyacrylamide sequencing gel is prepared for single-lane sequencing, full sequencing, and S_1 nuclease assays. These gels allow the resolution of single-stranded DNA ranging in length from 20 bp to 1 kb. A 5% polyacrylamide gel is used for S_1 analysis, and an 8% polyacrylamide gel is typically used for analyzing sequencing reactions.

TIME REQUIRED

2–3 hr

SPECIAL EQUIPMENT

Sequencing gel apparatus (e.g., BRL Model SO)[1]
Glass plates for conventional 34 × 40 cm gel
Spacers, shark's-tooth combs, and so on for 0.4-mm-thick gel
Gel-drying apparatus, large enough for sequencing gels (e.g., Bio-Rad)
Syringes, 10 and 60 ml

REAGENTS

Acrylamide (e.g., Biorad Ultrapure)
Bisacrylamide
Urea (e.g., Biorad Ultrapure)
Dichlorodimethylsilane, 5% in chloroform

---

1 Exact methods of sequencing gel apparatus setup may vary with the model used. See the instructions supplied with the apparatus.

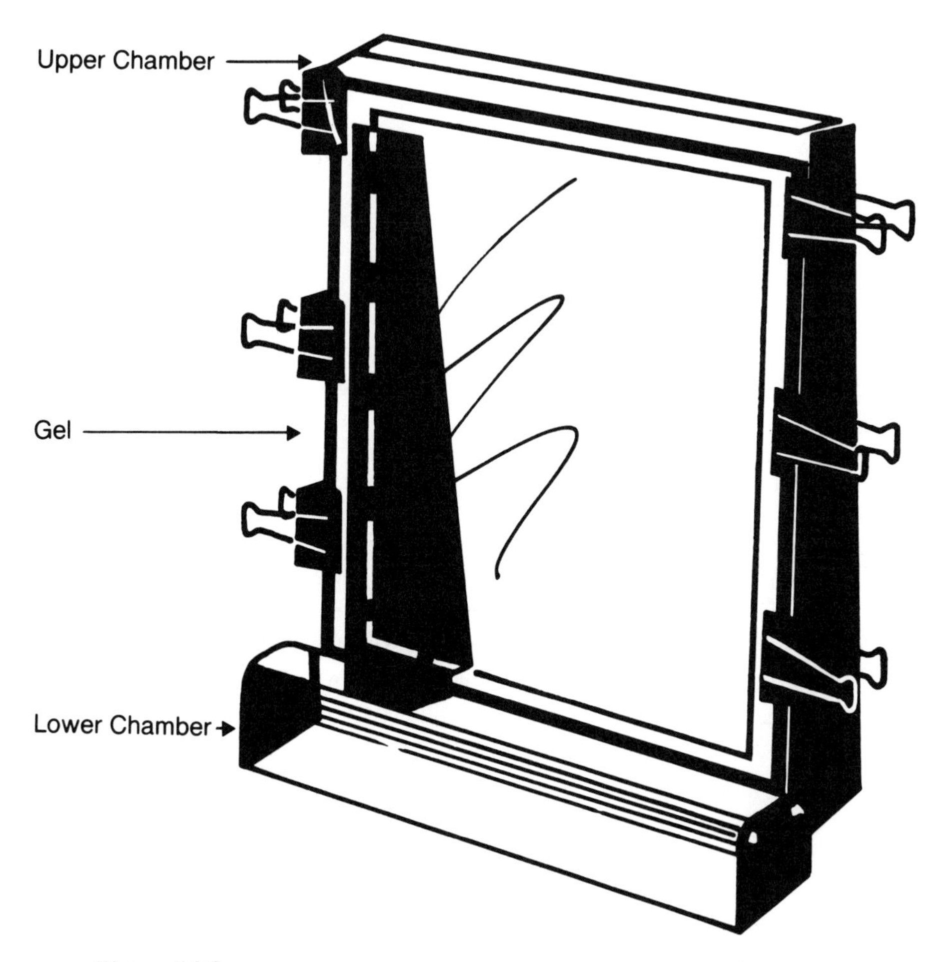

Figure 16.3
Diagram of a sequencing gel apparatus. Gel, between two glass plates, is attached to apparatus with clips. The open bottom of the gel is sitting in the lower buffer chamber, and the open top of the gel is contiguous with the upper buffer chamber. With gel clipped in place, both chambers are filled with running buffer. The samples are loaded into preformed wells in the top of the gel, the power supply is attached, and the gel is run.

Acrylamide solution (made in advance)

For 8% sequencing gel	For 5% S_1 gel
50 ml H_2O	53 ml H_2O
48 g urea	48 g urea
7.6 g acrylamide	4.75 acrylamide
0.4 g bisacrylamide	0.25 g bisacrylamide

Stir until uniformly in solution. Add 5 ml of 20× TBE buffer. Filter by vacuum through a 0.22-μm filter to remove any particles. Place solution in an Erlenmeyer vacuum flask and degas with vacuum for 5 min.

20× TBE buffer (and **1× TBE**)
0.8% agarose in **1× TBE**
TEMED (Biorad)
Ammonium persulfate, 10%. Stores for 2–3 days in dark at 4°C.
Sequencing loading solution

950 μl of deionized formamide
50 μl of *20× TBE buffer*
0.1 mg bromophenol blue
0.1 mg xylene cyanol

METHODS

Preparation of Plates

1. Clean glass plates (long front plate is 34 × 42 cm; short rear plate is 34 × 40 cm) with soap and water. Rinse with distilled H_2O and air-dry.
2. Silanize plates with 5% dichlorodimethylsilane in chloroform. Use gloves and work in fume hood to avoid breathing vapors. Add 2–3 ml of solution to plate and wipe thoroughly with paper towel until dry.
3. Place 1-cm-wide strips of Whatman 3MM paper along three edges of long glass plate, leaving one of the short edges free. Cover with shorter plate and clip plates together. Paper acts as spacers between the plates. Alternatively, 0.4-mm plastic spacers can be purchased. Large plate should extend at nonsealed end, which will become the top of the gel.
4. With a 10- or 12-ml syringe and a 20-gauge needle, inject 0.8% agarose solution (50°C) around the three edges with spacers to seal on three sides. Add more clips, being careful to place them directly over the spacers.
5. Stand plate upright, with unsealed side on top.
6. Place acrylamide solution (5% for S_1 nuclease assay, 8% for sequencing gels) in a 250-ml beaker.[2] Quickly add 20 μl of TEMED and 1 ml of 10% ammonium persulfate. Mix and quickly take up approximately 50 ml in a 60-ml syringe without a needle. Quickly inject solution into open-top end of

---

[2] A buffer gradient gel can be made instead of the single concentration sequencing gel. This will increase the effective resolution by decreasing the migration rate of smaller fragments at the bottom of the gel. Instead of making only the single (1× TBE) acrylamide solution, both 1× and 5× solutions are made. Make the 5× solution by combining 7.5 ml of H_2O, 12 g urea, 1.9 g acrylamide, 0.1 g bisacrylamide, and 6.25 ml of 20× TBE. Cool both solutions to 15°C on ice and add 20 μl of TEMED to each. Take up 12 ml of the 1× TBE solution in a 25-ml pipette carefully followed by 12 ml of the 5× TBE solution (in the same pipette). Allow an air bubble to go up through the solutions to gently mix. Empty this pipette into the gel plates. Fill the remaining plate volume with 1× TBE solution, as described. Proceed to step 7.

plates, being careful not to introduce air bubbles into the gel. Rock plates from side to side periodically to dislodge any air bubbles that may have formed while pouring in gel.[3]

7. When plate is almost full, lean gel over to an almost horizontal position and insert shark's-tooth comb into open end between the two plates with points up and the flat side into the gel mix. Depress comb tightly into gel to make an approximately 0.4-cm-deep valley in gel top. Clip tightly in place and allow gel to polymerize for 1–2 hr.[4]
8. After polymerization, plates with gel must be clipped into electrophoresis apparatus. If plastic spacers were used, bottom spacer must be removed. Turn over shark's-tooth comb so that teeth stick about 1 mm into top of gel in the valley previously formed. Do not push teeth too far into gel. The shorter plate will be sealed against the apparatus, with the longer plate in front. Flexible neoprene spacers are inserted between the plates and the apparatus to seal the top edge of the long plate against the top chamber. A more complete seal can be formed around the top by injecting hot agarose solution on and around the spacers.[5]
9. Add approximately 800 ml of 1× TBE buffer to the upper and lower chambers of the sequencing gel apparatus. Check for upper chamber leaks and correct if necessary. If air bubbles are trapped below the gel, remove them by injecting 1× TBE through a bent needle attached to a syringe.
10. Rinse spacers between teeth with 1× TBE buffer to dislodge urea diffusing out of the gel.
11. Load 2 μl of sequencing loading solution (without DNA or reactants) to alternate wells (spaces between teeth). Run gel at 50 W constant power until the dye front runs a few centimeters into the gel.[6] Also, check for leaks between wells; dye should only be in alternate lanes and should not leak into adjacent lanes.
12. Gel and apparatus are now ready for loading sequencing or S_1 reactions, which are then typically electrophoresed at 50 W constant power.[7]

---

[3] To provide an additional seal along the bottom, if a leak develops, mix 2 ml of acrylamide solution with 5 μl of TEMED and quickly inject to the site of the leak between plates to form a rapidly polymerizing strip of gel. Then inject the remaining solution into the gel frame, as described in step 6.

[4] This polymerization period is a convenient time to perform sequencing or single-lane sequencing reactions.

[5] Rinse valley thoroughly before inserting shark's-tooth comb.

[6] This setting recommended to maintain gel temperature during separation. Higher power settings may crack the plates.

[7] See notes 1 and 6.

SECTION **16-10.**

Sequencing M13 Clones

DESCRIPTION

This method allows the determination of the nucleotide sequence of inserts cloned into M13 vectors. A 17-mer oligonucleotide primer is annealed to the M13 vector in a position adjacent to the polylinker cloning region. This duplex is primer extended across the cloned insert using the large fragment of DNA polymerase I, also known as *Klenow polymerase.*

A small portion of the primer-extended products terminates at either G, A, T, or C bases in individual reaction tubes when a specific dideoxynucleotide in the primer extension reaction mix is incorporated into the newly synthesized strand, instead of the corresponding deoxynucleotide. By determining which of the four dideoxynucleotides generates a termination product at each position in the extended DNA chain, the nucleotide sequence can be determined. This dideoxy sequencing method is described in greater detail in Section 16-1. For a diagrammatic representation of this procedure, see Figure 16.1 on page 236. The procedure described here is a modification of that described in the BRL *M13 Cloning/Dideoxy Sequencing Manual* (see "References").

A number of M13 clones can be sequenced at the same time. Eight lanes of a sequencing gel are used per sample (i.e., the four reactions, electrophoresed for both a short and long time) to sequence approximately 300 bases past the primer. When BAL 31 deletion cloning is performed to determine longer sequences, as described in Section 16–3, the sequentially digested cloned inserts can be sequenced simultaneously to determine the whole sequence.

TIME REQUIRED

2 hr to run reactions

6 hr to run sequencing gels

2 hr to dry gels

1 to 2 days for autoradiography

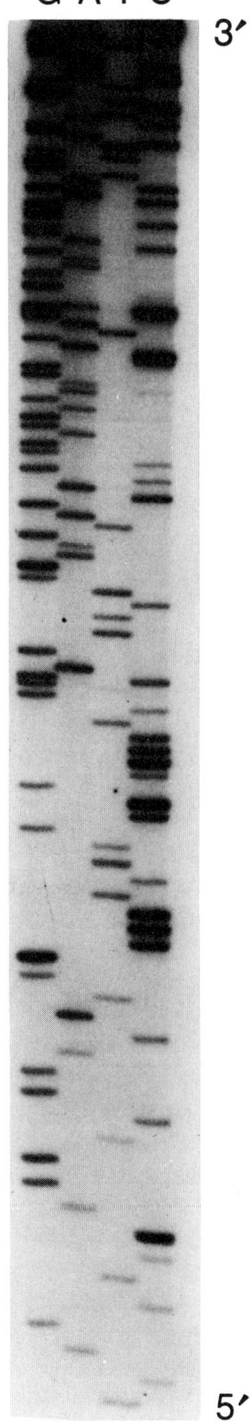

Figure 16.4
Photograph of an autoradiogram of a DNA sequencing gel.

SPECIAL EQUIPMENT

Sequencing gel apparatus, as described in Section 16-9
Vacuum gel dryer large enough for a 35 × 40 cm gel

REAGENTS

α-^{32}P-dATP (400 Ci/mmol) or α-^{35}S-dATP (1,000 Ci/mmol)[1,2]

M13 Universal Primer, 17-mer (e.g., Collaborative Research, "1000 Reactions", diluted to 250 μl with H_2O, stored at −20°C)

DNA Polymerase I, large fragment, Klenow, 5 U/μl, stored at −20°C (BRL)

10× Polymerase Reaction buffer, stored at −20°C

0.1 M DTT, stored at −20°C

dGTP, dATP, dTTP, and dCTP, diluted individually to 10 mM in H_2O and pH'd to 7.0 with NaOH. Store individually in 50-μl aliquots at −70°C for up to 1 year.

Make a working dilution of each dNTP stock to 2 mM with H_2O. Use these dilutions in the methods described below. Store the individual 2 mM dilutions at −20°C for up to 4 weeks.

ddGTP, ddATP, ddTTP, and ddCTP, each 10 mM, pH 7. Store individually in 5-μl aliquots at −70°C. (Concentrations can be confirmed as in Section 20-3.)

Make the following dilutions of one aliquot of each ddNTP:

ddGTP, 1 in 20 to 0.5 mM
ddATP, 1 in 40 to 0.25 mM[3]
ddTTP, 1 in 10 to 1.0 mM
ddCTP, 1 in 20 to 0.5 mM

These dilutions will be used in the methods described below.[4] Store at −20°C for up to 2–4 weeks.

---

1 Use proper care when handling radioactive material. Wear gloves throughout the procedure.

2 Either α-^{32}P or α-^{35}S-dATP may be used to label primer-extended strands in M13 sequencing. ^{35}S gives better resolution on gel autoradiograms but takes longer to expose film. In addition, ^{35}S is a lower-energy emitter and is somewhat safer to handle, in principle. The dilution of ddATP used is the major difference between the ^{35}S and ^{32}P methods.

3 See note 2.

4 The dilutions of the ddNTPs determine the rate of termination in this procedure. Higher ddNTP concentrations will result in shorter average chains, lower concentrations result in longer chains. The concentration of ddNTPs may be varied to improve the quality of an "underterminated" or "overterminated" sequence. The values given here are average values that have given a good range of terminations for determining the first 300 bases of an insert.

ddATP, 0.06 mM, pH 7 (1 in 160 dilution of 10 mM stock), used for ^{35}S-dATP sequencing.[5]

Sequencing gel loading solution
- 950 μl deionized formamide
- 50 μl *20× TBE buffer*
- 0.1 mg bromophenol blue
- 0.1 mg xylene cyanol

METHODS

In Advance

Prepare 8% polyacrylamide, 8 M urea sequencing gel, as described in Section 16-9. Prepare 100 μl chase solution by combining 25 μl each of 2 mM dilutions of dGTP, dATP, dTTP, and dCTP (store at −20°C for up to 4 weeks).

Prepare G°, A°, T°, and C° solutions from 2mM stocks of:

	dGTP	dATP	dTTP	dCTP
G°	1 μl	—	20 μl	20 μl
A°	20 μl	—	20 μl	20 μl
T°	20 μl	—	1 μl	20 μl
C°	20 μl	—	20 μl	1 μl

Store at −20°C for up to 4 weeks.

Set Up Reactions

1. Make 9.5 μl of annealing mix for each sample to be run as follows:

DNA sample, from Section 16-7	5 μl
Primer, 17-mer	1 μl
10× polymerase reaction buffer	1.5 μl
H_2O	2 μl

Heat mixture to 55°C for 5 min. Cool over 1–2 hr to room temperature (25°C) in a water bath to allow annealing of primer to M13 DNA.

2. For each sample to be run make 5 μl of polymerase mix:

α-^{32}P-dATP or α-^{35}S-dATP	3 μl
DNA polymerase I, large fragment	1 μl
0.1 M DTT	1 μl

---

5 See note 2.

3. Add 9.5 μl of annealing mix to 5 μl of polymerase mix to form 14.5 μl of primer template mix.
4. For each M13 clone to be sequenced, set up four reaction tubes, labeled G, A, T, and C, containing the following:

G tube	A tube	T tube	C tube
1 μl G°	1 μl A°	1 μl T°	1 μl C°
1 μl ddGTP	1 μl ddATP	1 μl ddTTP	1 μl ddCTP

Note: use the four ddNTP dilutions indicated in the "Reagents" section. If α-^{35}S-dATP is used instead of ^{32}P, use 1 μl of 0.06 mM dATP in the A tube instead of 1 μl of the 0.25 mM dilution.

5. To each of the above four tubes, add 3 μl of the 14.5-μl primer template mix from step 3.

Primer Extension

6. Incubate tubes for 10 min at room temperature.
7. Add 1 μl of chase solution to each tube. Incubate for 10 min at room temperature.

Run Samples on Gel

8. Add 4 μl of sequencing gel loading solution to each tube. Incubate for 1 min in a boiling water bath to denature primer-extended chains.
9. Load 2 μl from each tube on individual lanes of 8% acrylamide sequencing gel, as described in Section 16-9. Load G, A, T, and C reaction samples from each M13 clone to be sequenced in four adjacent lanes of the gel. Asymmetrical loading of sample sets of the gel is useful for later orientation.
10. Run gel for approximately 2 hr at 50 W constant power until lower dye (bromophenol blue) is near the bottom of the gel. This resolves the first 100–150 bp (short electrophoresis).
11. A second set of 2-μl aliquots from the same four tubes is loaded and run on a sequencing gel for 5 hr until the xylene cyanol has migrated 60–70 cm. This resolves nucleotides from 150 to 300 bp (long electrophoresis).

    Alternatively, both the long and short electrophoresis can be run on the same gel, if there is room. To do this, run the long samples for 3.5–4 hr until the xylene cyanol dye front is at the bottom of the gel. Reboil samples from step 8 for another minute and apply a second set of 2-μl samples (short) in four adjacent lanes of the same gel. Run gel an additional 2 hr to move the new xylene cyanol dye front of short sample 20 cm down the gel.

12. When gel is completely electrophoresed, turn off power supply and carefully unclip and remove glass plate. The gel usually adheres firmly to one plate.[6,7]

13. Cut out a piece of Whatman 3MM filter paper slightly larger than the gel and press firmly against the gel.

14. Carefully flip over the paper/gel/glass so that glass plate is on top. Remove top glass. Gel is now stuck to 3MM paper support. Cover gel with plastic wrap and place in a vacuum gel dryer. Dry gel under vacuum at 80°C for 1–2 hr, depending on the vacuum.

15. After gel is dry, the plastic wrap should be stripped away from it before autoradiography if ^{35}S-dATP was used. The plastic wrap may be left in place if ^{32}P-dATP was used.

16. Determine orientation of gel relative to sample loading and label. Autoradiograph with XAR-5 film, gel side toward film. Expose overnight without screen at room temperature. Develop film.

17. Sequences of the inverse complementary strand to the cloned DNA can be read from the bottom (nearest the primer), starting with the lanes run in steps 9 and 10. Longer primer-extended species are better resolved in the first four lanes from step 11. Overlapping sequences, resolved in both sets of sample electrophoresis, will indicate how to combine the two readings.

REFERENCES

M13 Cloning/Dideoxy Sequencing Manual, BRL, Gaithersburg, Md., 1980.

Sanger, F., Nicklen, S., and Coulson, A. R., Proc. Natl. Acad. Sci., USA *74*:5463, 1977.

---

[6] An edge of a forceps can be used to pry the plates carefully apart off the gel. Put plate with the gel adhering on bottom and slowly remove top plate. Note the original orientation of samples, relative to which plate the gel is adhering to, for future reference.

[7] If ^{35}S was used in the sequencing reactions, the gel can be freed of urea, which may quench the signal. Carefully place the gel and plate in a 40 × 50 × 5 cm tank and cover with 1 liter of 5% methanol/5% acetic acid. Incubate for 20–30 min for urea to diffuse out of gel. Be careful not to dislodge gel from the plate. Syphon off liquid and continue with step 13. For additional information on ^{35}S sequencing, see: Williams, S. A. et al., BioTechniques *4*:138, 1986, Ornstein, D. L. and Kashdan, M. A., BioTechniques *3*:476, 1985.

SECTION

17

Further Characterization of Cloned DNA

SECTION

17-1. S_1 Nuclease Protection Assay

DESCRIPTION

This assay allows the precise identification of pertinent gene boundaries in RNA transcripts. These can include intron/exon junctions and the 5′ or 3′ ends of transcripts. Further, this assay allows a specific comparison between the RNA and a labeled DNA probe. This is a more stringent and sensitive method than Northern blot analysis, but is also more tedious and requires more knowledge of the structure of the gene being expressed and analyzed.

The S_1 nuclease assay works by endonuclease digestion of single-stranded DNA. Thus, if the ^{32}P-DNA does not base pair precisely to the hybridizing RNA, extra tails or loops not protected by RNA hybridization will be excised. Resolution of the undigested S_1-protected species yields detailed information about the regions of sequence homology between the probe and the protecting mRNA species.

There are eight steps to this procedure as follows:

1. Start with an M13 clone whose primer-extended product will be complementary to the mRNA of interest, and total RNA or poly(A^+)RNA from tissue samples or cells to be tested.
2. Make radiolabeled complementary copy by primer extension of M13 clone, starting at the universal priming site.
3. Cut beyond the cloned insert in the primer-extended probe with a suitable RE.
4. Run on denaturing agarose or acrylamide gel to isolate the probe.
5. Electroelute probe from gel.
6. Hybridize probe to RNA samples under high-stringency conditions that allow only RNA–DNA duplexes to form.
7. Treat with S_1 nuclease; nonhybridized portions of the probe will be digested.
8. Denature and run on high-resolution denaturing gel to visualize and determine the size of protected hybridized species.

TIME REQUIRED

Day 1—8 hr to prepare probe for hybridization (steps 1 to 6)
Day 2—4–5 hr to do S_1 nuclease reaction and run on gel (steps 7, 8)
Day 3—1 hr to analyze gel

SPECIAL EQUIPMENT

Minigel apparatus
Sequencing gel apparatus and plates
Scintillation counter
Gel dryer

REAGENTS

Primer: 17-mer universal primer (e.g., Collaborative Research)
10× polymerase buffer
α-^{32}P-dATP, 400 Ci/mmol (NEN, Amersham)
A° mix: 20 μl each of dGTP, dTTP, and dCTP from 2 mM stocks (Section 16-10)
DNA polymerase I, large fragment, Klenow, 5 U/μl (BRL)
Chase solution, 0.5 mM each dATP, dTTP, dCTP, and dGTP (same as in M13 sequencing, Section 16-10)
RE and RE buffer (depends on the clone used)
10× alkaline denaturing gel buffer
 300 mM NaOH
 20 mM EDTA

Alkaline gel loading buffer
 0.3% bromocreosol green
 50% glycerol
 50% 10× alkaline denaturing gel buffer

1× TBE buffer
SS-phenol
Chloroform
3 M sodium acetate, pH 7
Ethanol
Hybridization buffer (10×)
 200 mM Tris, pH 7.4
 4 M NaCl
 10 mM EDTA

0.1 M DTT
10% SDS
Formamide (deionized)
Salmon sperm DNA, 2 mg/ml; denature by boiling for 5 min. Store at 4°C.
S_1 nuclease, reconstituted from powder at 33 U/μl (BM); will last up to 1 month
S_1 buffer (5×)

3 ml *5 M NaCl*
1.66 ml 100 mM $ZnSO_4$
1 ml *3 M sodium acetate, pH 4.5*
4.33 ml H_2O

tRNA from yeast, 10 mg/ml
Gel loading solution

850 μl formamide (deionized)
50 μl *20× TBE buffer*
50 μl *xylene cyanol*
50 μl *bromophenol blue*

TE buffer
pBR322
*Eco*RI
*Hin*fI

METHODS

In Advance

Prepare M13 clone with inserted DNA of interest; sequence of the inserted DNA should be known and will generate a probe complementary to RNA of interest when primer extended. Wear gloves and use sterile or autoclaved plasticware throughout this method. Prepare RNA samples to be hybridized with probe (10–30 μg per tube, dried in a vacuum centrifuge). If poly(A^+) RNA is used, add 2 μg per tube. Prepare ^{32}P-labeled size markers for acrylamide gel (see note 4).

Primer Extension

1. For each reaction, in a 1.5-ml microfuge tube combine:

   1 μl primer
   1.5 μl polymerization buffer
   2.5 μl H_2O

2. Add 5 μl of M13 DNA (prepared as in Section 16-7). Mix. Briefly spin in microcentrifuge to concentrate liquid on bottom. Place sample in a 55°C

water bath. Cool slowly by removing water bath from heat and allowing bath and tube to reach room temperature. This usually takes 1 hr.

3. Add, in order, the following solutions to initiate primer extension reaction:

4 μl of α-^{32}P-dATP[1]
4 μl of A° mix
1 μl of 0.1 M DTT
1 μl of DNA polymerase I, large fragment (5 U/μl)

Briefly spin in microcentrifuge to bring components to bottom.

4. Incubate for 10 min at room temperature.

5. Add 2 μl of chase solution.

6. Incubate for an additional 10 min at room temperature.

7. Extract DNA with SS-phenol/chloroform and precipitate with ethanol, as described in Section 20-1. Dissolve pellet in proper RE buffer. Cut primer-extended M13 clone with RE (Section 5-4). Select an RE that will make a single cut in the vector beyond the portion of the insert expected to hybridize mRNA. Do not cut insert (see Figure 17.1). Use a total volume of 20 μl for the RE digest (Section 5-4).

8. After RE digestion, add 5 μl of alkaline gel loading buffer. Place on ice.

Probe Purification: Alkaline Denaturing Agarose Gel[2]

9. Prepare gel as follows: Boil 22.5 ml of H_2O with 0.25 g of agarose in a 125-ml Erlenmeyer flask in a microwave oven (or boiling water bath) for 2 min. Add 2.5 ml of 10× denaturing gel buffer to agarose solution. Increase volume to 25 ml with H_2O, if necessary. Mix by swirling. Place 14 ml on minigel plate with 0.85-cm toothed comb (see Section 9-1 for minigel procedure). After allowing gel to harden for 30 min, place in a minigel apparatus with 500 ml of 1× alkaline denaturing gel buffer.

10. Boil sample from step 8 for 2 min.

11. After gel is ready, load boiled sample on gel (25 μl).

12. Run gel at 50 V until dye has run approximately halfway (1–1.5 hr). The gel apparatus is now radioactive; handle appropriately.

13. Remove gel from apparatus and cover gel with clear plastic wrap (Saran Wrap works well). In a darkroom, expose the wrapped gel to X-ray film, for 30 sec to a few min.[3] Mark the position of the gel on the film to allow

---

[1] Handle radioactive reactants and waste properly. Wear gloves for this procedure.

[2] Denaturing 8 M urea polyacrylamide gels have been used for purifying the probe. The alkaline agarose gel is faster and easier to handle. The polyacrylamide gel may be needed for probes of very small length (i.e., fewer than 300 bases). See Section 16-9, but use 1-mm spacers.

[3] Align corner of film with gel to determine position of bands. Place glass plate over film during exposure to hold steady.

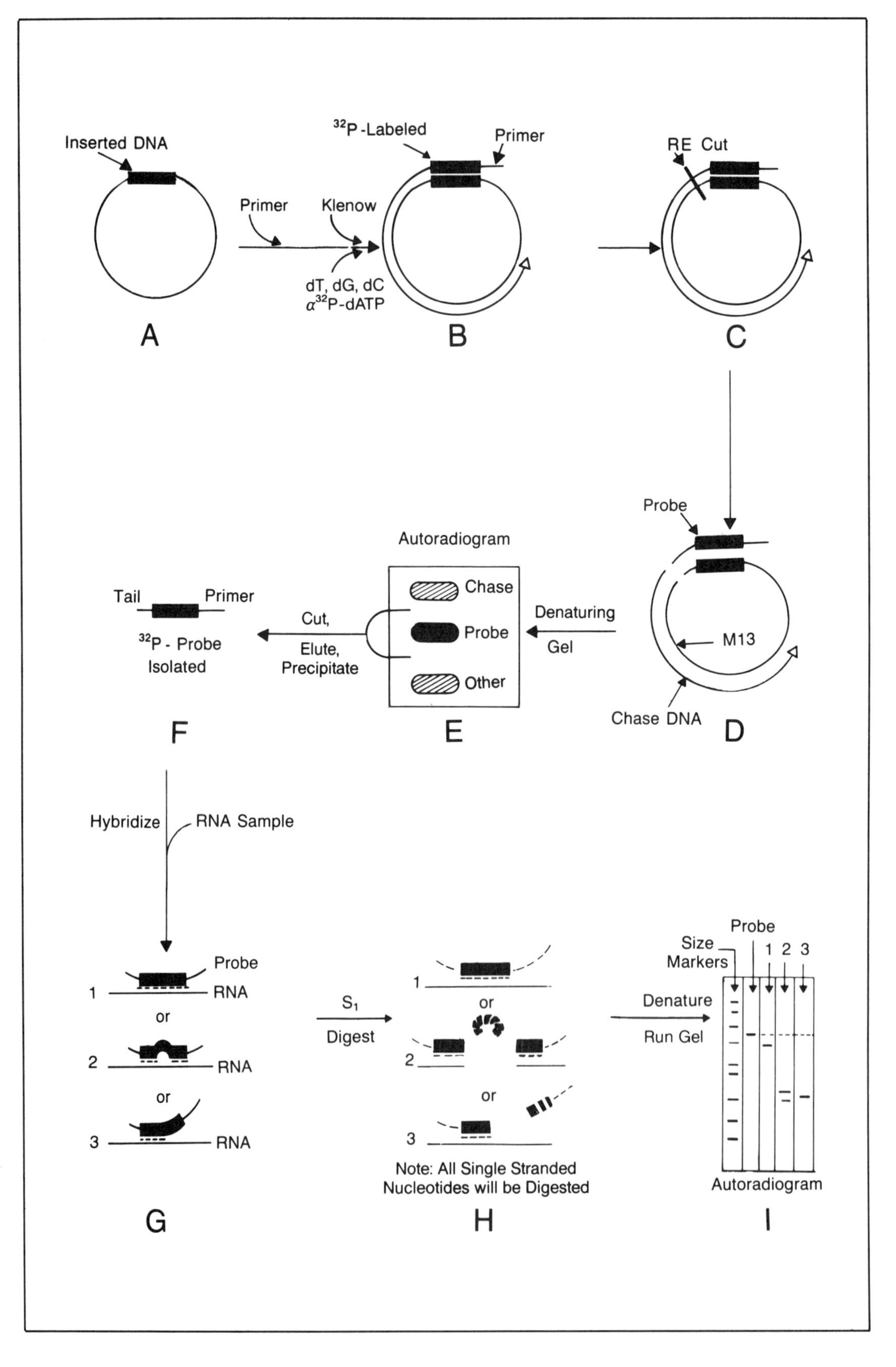
Inserted DNA
A
Primer
Klenow
dT, dG, dC
α32P-dATP
32P-Labeled
Primer
B
RE Cut
C
Probe
M13
Chase DNA
D
Denaturing
Gel
Autoradiogram
Chase
Probe
Other
E
Cut,
Elute,
Precipitate
Tail
Primer
32P - Probe
Isolated
F
Hybridize
RNA Sample
Probe
1
RNA
or
2
RNA
or
3
RNA
G
S1
Digest
1
or
2
or
3
Note: All Single Stranded
Nucleotides will be Digested
H
Denature
Run Gel
Probe
Size
Markers
1 2 3
Autoradiogram
I

alignment of the gel with the autoradiogram after exposure and development.

14. Superimpose developed film over gel and cut out the appropriate primer-extended DNA insert with an unused razor blade. Typically, the probe will appear as the middle of three major autoradiographic bands (see illustration). If the position of the desired band is in question, it is helpful to run molecular weight markers on the gel to help determine the size of the correct band. A convenient set of molecular weight markers for S_1 probes less than 1 kb in size can be prepared.[4]

Elution of Probe

15. Place gel piece in dialysis bag with 200 μl of 1× TBE buffer. Electroelute as described in Section 9-3. Probe will remain inside bag.[5]

---

[4] *Hin*fI- and *Eco*RI-cut pBR322 markers are convenient for many S_1 nuclease experiments. To generate these ^{32}P markers, mix:

1 μg pBR322 DNA
2 μl buffer: **600 mM NaCl**
60 mM Tris, pH 7.4
60 mM $MgCl_2$
60 mM β-mercaptoethanol
4 μl α-^{32}P-ATP (high or low specific activity)
0.5 μl *Hin*fI, 10 U/μl
0.5 μl *Eco*RI, 10 U/μl
12 μl H_2O

Incubate for 30 min at 37°C. Add 0.5 μl of DNA polymerase I, large fragment (Klenow) to fill in ends. Incubate for 15 min at 37°C. Incubate for 10 min at 65°C to inactivate enzyme. Add 1 μl of a saturated solution of blue dextran. Separate unincorporated triphosphates from incorporated counts by chromatography using a 5-ml G50 or BioRad P10 column in TE buffer. Collect only the fractions that run with blue dextran marker (the void volume of the column). Size markers will be 998, 634, 517/506, 396, 344, 298, 221, 220, 154, and 75 bases.

[5] See note 1.

Figure 17.1
Schematic diagram of S_1 Nuclease protection assay. (**A**) M13 vector with inserted DNA of interest is isolated. (**B**) A second strand, with radiolabelled nucleotides, is synthesised extending from the primer region. (**C**) The double-stranded product is cut with an appropriate RE, distal to the insert region. (**D**) The cut product is isolated on the denaturing gel. (**E**) From the autoradiographed gel, the radiolabelled probe, complementary to the DNA of interest, is identified. (**F**) The probe is isolated from the gel. (**G**) The probe is hybridized with RNA sample. (Note that at least 3 possible situations exist that would result in different bands on the gel.) (**H**) The hybridized product is subjected to S_1 Nuclease digestion; all single-stranded nucleotides will be digested. (**I**) The sample is run on a denaturing polyacrylamide gel to determine the size of nondigested ("protected") hybridized pieces.

16. Transfer eluted probe in TBE buffer from bag to a microfuge tube. Rinse bag with an additional 200 μl of TBE buffer and combine samples. Spin sample in microcentrifuge for 30 sec to remove any transferred gel pieces. Transfer supernatant to a new tube.
17. Purify probe by adding 1 volume of SS-phenol and 1 volume of chloroform. Mix well. Spin for 1 min in microcentrifuge. Remove upper aqueous phase to new tube.
18. Add 1/10th of 1 volume (40 μl) of 3 M sodium acetate, 2 μl of tRNA, and 2.5 volumes (1 ml) of ethanol. Mix well. Freeze on dry ice for 10 min. Spin in microcentrifuge for 5 min at 4°C. Decant supernatant. Rinse pellet in 200 μl of ethanol. Spin in a microcentrifuge for 5 min at 4°C. Carefully remove supernatant and dry pellet under vacuum.
19. Make 3.5 ml of the following buffer:

    3 ml deionized formamide
    0.4 ml 10× hybridization buffer
    40 μl 10% SDS
    60 μl H_2O

20. Dissolve DNA pellet from step 18 in 300 μl of buffer from step 19.
21. Count 5 μl of this sample with 5 ml scintillation fluid in liquid scintillation counter. Adjust sample volume with buffer from step 19 so that there will be 25,000–100,000 cpm per 35 μl.[6]

Hybridization of Probe to RNA

22. Prepare each RNA sample to be analyzed by ethanol precipitation in a microfuge tube. Rinse the pellet with 80% ethanol and dry the pellet under vacuum (Section 20-1). To each RNA sample, stored dried in 1.5-ml microfuge tubes, add 35 μl of probe (step 21) and 5μl of H_2O. Add one tube with 10 μg of tRNA as the control. Mix by vortexing to dissolve RNA. Briefly centrifuge all liquid to bottom.
23. Heat tubes for 15 min at 75°C.
24. Determine correct incubation temperature for hybridization.[7] Estimates can be made from sequence of probe by determining percentage of G and C bases and using the following approximate temperatures:

---

6 If there are fewer than 25,000 cpm per 35 μl, probe may not be usable. More likely, probe counts will be greater than 100,000 cpm per 35 μl and must be adjusted accordingly.

7 These temperatures are empirically derived and are calculated to be about 5–6°C above the DNA–DNA melting temperature but below the RNA–DNA melting temperature in high-formamide buffers. To optimize the assay, it is advisable to run the hybridization reaction the first time a probe is used at various temperatures near the predicted optimal value to empirically determine the best assay conditions.

40% GC→45°C
50% GC→50°C
60% GC→55°C

25. Incubate tubes overnight at the determined incubation temperature.

Next Day: S_1 Nuclease Reaction

26. Prepare the following mixture (multiply volumes by the number of S_1 reaction samples):

    1 μl salmon sperm DNA
    1.5 μl S_1 nuclease
    280 μl H_2O
    80 μl 5× S_1 buffer

    The amount of S_1 nuclease may have to be adjusted empirically for optimal results.

27. Remove each tube *individually* from incubation bath and *immediately* add 350 μl of the above mixture.[8] Mix well by pipetting several times in and out of the tube. Use a new pipette tip for each tube.

28. Incubate tubes for 1 hr at 37°C. During incubation, prepare an 8 M urea, 5% polyacrylamide sequencing gel, as described in Section 16-9.

Purify Sample

29. Add 400 μl of SS-phenol and 400 μl of chloroform. Mix well. Spin in microcentrifuge for 3 min. Transfer upper aqueous phase to a new microfuge tube.

30. Add 1/10th of 1 volume (40 μl) of 3 M sodium acetate, pH 7, and 1 ml of ethanol. Mix well. Freeze on dry ice for 10 min. Spin in microcentrifuge for 5 min at 4°C.

31. Add 400 μl of 80% ethanol to wash pellet. Spin in microcentrifuge for 3 min. Decant ethanol.

32. Repeat ethanol wash, as in step 31. Dry pellet under vacuum.

Run Sequencing Gel

33. Redissolve pellet in 3 μl of TE buffer. Add 4 μl of gel loading solution. Vortex to resuspend the pellet.

34. Boil samples for 2 min to denature.

---

[8] It is important to add mix quickly to each tube while still at or near incubation temperature.

35. Load 4 μl of sample on 5% sequencing gel and run, as described in Section 16-9, for approximately 3 hr at 50 W. Also include the sample containing yeast tRNA (negative control), a lane with 10 μl of untreated probe (step 21), and a lane with 4 μl of size markers denatured in gel loading solution in a manner identical to that of the samples (steps 33 and 34). The xylene cyanol dye (light blue) runs with approximately 150 bases; run gel an appropriate length of time to resolve the expected S_1-digested species.

36. When gel electrophoresis is completed, carefully separate plates. Press a sheet of Whatman 3MM paper against the gel. Roll to smooth.

37. Carefully invert gel (paper side down) and gently lift off the remaining plate.

38. Cover gel in clear plastic wrap, and trim off excess 3MM paper.

39. Dry gel and paper on a vacuum gel dryer at 80°C for 1–2 hr.

40. Autoradiograph the gel overnight at −70°C with intensifying screen and adjust the time of exposure, based on the initial overnight film, to make a second exposure, if necessary.

Analyzing Gel

41. If the probe is hybridized to RNA in samples, it will appear as a distinct band, somewhat smaller than the nonhybridized probe length, because the RE cleavage was distal to the cloned sequence in M13 DNA and a 17-mer primer was extended to make the probe. If the probe was hybridized with RNA containing an internal noncomplementary region of five or more bases, the S_1 nuclease will digest these unpaired bases. This digested heteroduplex will result in smaller S_1-protected bands on the autoradiogram. A band representing DNA–DNA duplex formation between the full length of the probe and a small amount of contaminating unlabeled M13 DNA template usually present will show up on all lanes independent of the RNA present during hybridization. RNA species specific to the sample hybridization will show up in sample lanes and not in the tRNA hybridization lane, which serves as a negative control sample.

REFERENCES

Berk, A. J., and Sharp, P. A., Cell *12:*721, 1977.

Casey, J., and Davidson, N., Nucleic Acids Res. *4:*1539, 1977.

Ley, T., Anagnou, N., Pepe, G., and Nienhaus, A., Proc. Natl. Acad. Sci., USA *79:*4775, 1982.

Vogt, V. M., Meth. Enzymol. *65:*248, 1980.

SECTION

18

Transfection of Mammalian Cells in Culture

SECTION **18-1.**

Calcium Phosphate Transfection of Nonadherent and Adherent Cells with Purified Plasmids*

DESCRIPTION

This is an efficient method for the introduction of cloned genes on recombinant plasmids into mammalian cells. There are slightly different protocols for use with nonadherent and adherent cells.[1]

TIME REQUIRED

4–6 hr plus cell growth time

SPECIAL EQUIPMENT

Cell culture equipment (incubator, biohood, automatic pipettor, etc.)

REAGENTS

2× HBSP buffer[2]

- 1.5 mM Na_2HPO_4
- 10 mM KCl
- **280 mM NaCl**
- 12 mM glucose
- 50 mM HEPES, pH 7

---

* This method was contributed by Dr. Shoshana Segal.

1 Examples of nonadherent cells are mouse B cell lines, including MPC11, 7OZ, 1881, and so on. Examples of adherent cells are L cells, 3T3, CV-1, etc.

2 Check to see if 2× HBSP buffer will form a precipitate before actually attempting a DNA transformation. To do this, perform steps 1 and 2 of this method without DNA. After 15 min of incubation in the cell culture hood, an opaque precipitate should appear.

Make 100 ml of 2× solution. Pass through 0.22-μm filter.
Store in 5-ml aliquots at −20°C.

15% glycerol in 1× HBSP buffer. Make fresh.

2 M $CaCl_2$. Store at −20°C.

Phosphate-buffered saline (PBS)

Cell culture growth medium; depends on cells used.

METHODS

In Advance

Test all cell lines to be used in transfection for sensitivity to glycerol.[3] Prepare and purify plasmid for transfection. Sterilize all buffers through 0.22-μm filters.

To Transfect Nonadherent Cells (for adherent cells, skip to "Transfection of Adherent Cells with $CaPO_4$")

1. Prepare $CaPO_4$/DNA mixture by combining in order:

   H_2O (to bring final volume to 1 ml)
   Plasmid DNA (10–25 μg)
   2 M $CaCl_2$, 62 μl (0.125 M final)
   2× HBSP, 500 μl

   Final volume is 1 ml. To make precipitate, place H_2O in the bottom of a sterile 15-ml conical tube. Gently add DNA to the H_2O. *Do not mix.* Add 2 M $CaCl_2$. *Do not mix.* Use a 1-ml pipette attached to an automatic pipettor (e.g., Pipet-aid) to add the 500 μl of 2× HBSP buffer. Gently place pipette inside the tube, touching the bottom. Slowly release the buffer and blow in about five bubbles with the pipettor to achieve a gentle mixing of ingredients.[4]
2. Allow precipitate to form for 15 min in a cell culture hood at room temperature. Solution should appear opaque or slightly cloudy.
3. During this time, collect cells by centrifugation. Use 1×10^7 cells per transfection (for 10–25 μg of plasmid DNA). Spin cells at 500 × *g* for 5 min.
4. Discard supernatant. Wash cell pellet once with 1 ml PBS. Recentrifuge to pellet cells.

---

[3] To perform a glycerol sensitivity test:
  a. Add 15% glycerol in 1× HBSP to actively growing cells.
  b. Incubate at 37°C for different times up to 3 min. Examine under microscope.
  c. If cells round up and die during incubation, do not glycerol shock after transfection.

[4] This gentle mixing will prevent the formation of large precipitates that decrease the efficiency of transfection.

5. Resuspend cells in 25 μl of 1× HBSP.
6. Add $CaPO_4$/DNA precipitate to the cells and pipette up and down once.
7. Let stand for 15 min at room temperature in a cell culture hood.
8. Add 5 ml of complete medium.[5]
9. Plate in T-25 flasks or 6-cm tissue culture dishes.
10. Incubate for 4 hr at 37°C in a humidified CO_2 incubator.[6]
11. Collect cells by centrifugation at 500 × *g* for 5 min.
12. Discard supernatant. Resuspend pellet in 5 ml of PBS to wash cells. Spin as in step 11.
13. At this point, if cells are resistant to glycerol, do a glycerol shock as described below. Otherwise, do not shock cells with glycerol.

Glycerol Shock

14. Resuspend cells in 1 ml of 15% glycerol in 1× HBSP.
15. Incubate for 1–3 min at 37°C (depending on cell line).
16. Remove glycerol by centrifugation at 500 × *g* for 5 min. Discard supernatant.
17. Add 5 ml of PBS, centrifuge for 5 min at 500 × *g*, and discard supernatant.
18. Resuspend cell pellet in 10 ml of complete growth medium[7] (with serum, glutamine, antibiotics). Plate in 10-cm tissue culture dishes.
19. Incubate for 36–48 hr at 37°C in a humidified CO_2 incubator.
20. If necessary, select for stable transfectants. Section 18-4 describes one method that selects for resistance to G418.

Transfection of Adherent Cells with $CaPO_4$

In Advance

From 16 to 20 hr before transfection, remove cells from adherent growth with trypsin, inactivate with trypsin inhibitor,[8] and plate in T-25 flasks or 6-cm tissue culture dishes at 3×10^5 cells per dish in complete medium.[9] From 2 to 4 hr prior to transfection, remove medium from plates, add 5 ml of complete growth medium, and incubate at 37°C.

---

[5] Complete medium contains serum, 2 mM glutamine, and antibiotics. The percentage of serum included depends on the cell line.

[6] CO_2 levels should be 5–7%.

[7] See note 5.

[8] Trypsin inhibitor is present in fetal calf or calf serum. To inactivate trypsin, add a volume of serum equal to the volume of trypsin used.

[9] See note 5.

21. Prepare $CaPO_4$/DNA as in steps 1 and 2, but cut all volumes in half (total, 0.5 ml).

22. Let mixture stand in cell culture hood for 30 min at room temperature.

23. Mix precipitate by pipetting up and down once. Add very slowly to plated cells. Do not mix vigorously. Gently rock plate a few times. Medium should turn yellow and become turbid.

24. Incubate for 4 hr at 37°C in a humidified CO_2 incubator. After first 10 min of incubation, remove from incubator and examine dish under microscope. A fine precipitate should be seen on top of the cells.

25. Glycerol shock (steps 14 to 20), if cells will tolerate.[10]

26. To shock cells, remove growth medium and wash once with 5 ml of PBS. Remove PBS, add 1.2 ml of 15% glycerol in 1× HBSP, and incubate for 1–3 min at 37°C.

27. Remove glycerol (if used) and wash once with 5 ml of PBS. Remove PBS and add 5 ml of complete growth medium.

28. Incubate for 36–48 hr in a humidified CO_2 incubator.[11]

29. Plate cells in a selective medium after trypsinization if selection is necessary for stable transfectants.

30. For transient expression, harvest cells after 36–48 hr and analyze RNA, DNA, or protein.

REFERENCES

Gorman, C., Moffat, L., and Howard, B., Molec. Cell. Biol. *2:*1044, 1982.

Graham, F., and Van der Eb, A., Virology *52:*456, 1973.

Parker, B., and Stark, G., J. Virology *31:* 360, 1979.

---

[10] See note 3.

[11] See note 6.

SECTION **18-2.**

DEAE Dextran–Mediated Transfection of Nonadherent and Adherent Mammalian Cells*

DESCRIPTION

The DEAE dextran method for introducing cloned genes into mammalian cells is an alternative to the $CaPO_4$ technique for transformation. It has been observed that this method is much better for studying the transient expression of transfected genes. In addition, it works better than $CaPO_4$ for some cell lines and is somewhat simpler to perform. Some cells, however, yield better results with $CaPO_4$.

The methods differ for nonadherent and adherent cells.[1]

TIME REQUIRED

2 hr

REAGENTS

Diethylaminoethyl Dextran DEAE Dextran, 500,000 M.W. (GIBCO)

1 M Tris, pH 7.4

Cell culture growth medium for cells to be transfected

---

* This method was contributed by Dr. Shoshana Segal.

[1] Cells differ in their ability to withstand DEAE dextran medium. Some cells cannot tolerate more than 1 hr in this medium. Test individual cell lines for their ability to withstand DEAE-dextran and adjust incubation time from 1 to 3 hr, accordingly.

Plasmid DNA to transfect cells
Mammalian cells for transfection[2]

METHODS

In Advance

Prepare cell growth medium containing 250 μg/ml DEAE dextran and 50 mM Tris, pH 7.4, by adding 0.125 g DEAE dextran and 25 ml of 1 M Tris, pH 7.4, to 500 ml of serum-free growth medium. Shake for 30 min at 37°C. Sterilize through 0.22-μm filter.

Transfection of Nonadherent Cells

1. In a 15-ml conical sterile tube, collect cells by centrifugation at 500 × *g* for 5 min (use 1×10^7 cells per transformation). Resuspend cells in 5 ml serum-free growth medium to wash cells. Mix gently and centrifuge for 3 min at 500 × *g*. Decant medium.
2. Resuspend cells in 1 ml of growth medium containing DEAE dextran (see "In Advance") and add 10–25 μg of plasmid DNA to the cells.
3. Incubate for 1 hr at 37°C. Bring up and down with a pipette once to mix every 15 min.
4. Add 9 ml of growth medium (without serum, glutamine or antibiotics) to wash cells. Mix gently. Centrifuge at 500 × *g* for 5 min to collect cells. Discard supernatant. Repeat this wash one more time.
5. Resuspend cells in 10 ml of complete growth medium (containing serum, 2 mM glutamine, and antibiotics). Plate in 10 cm tissue culture dishes.
6. Incubate for 44–48 hr. at 37°C.
7. Harvest cells and analyze for transient expression of transfected gene.
8. If necessary, plate harvested cells in selective medium to establish stable transfected clones.

Transfection of Adherent Cells

9. From 16 to 20 hr before transfection, plate cells in 10-cm dishes at a density of 1×10^6 cells per dish in 10 ml of complete growth medium. Incubate for 16–20 hr at 37°C.
10. Wash cells twice by replacing medium with 10 ml of serum-free growth medium.
11. Add 5 ml of DEAE dextran medium (see "In Advance") containing 10–25 μg of plasmid DNA.
12. Incubate for 1–3 hr, depending on the cell line, at 37°C in a CO_2 incubator.

---

[2] Cells that have the potential to differentiate in culture (e.g., F9 teratocarcinoma) need to be tested for induction of differentiation by DEAE dextran.

13. Wash cells twice with 10 ml of serum-free medium.
14. Add 10 ml of complete medium.
15. Incubate cells for 48 hr at 37°C in a CO_2 incubator.
16. Harvest cells for analysis, if transient expression is to be examined.
17. If necessary, replate cells in a selective medium to obtain stable transfectants.

REFERENCES

Sompayrac, L., and Danna, K., Proc. Natl. Acad. Sci., USA *78:*7575, 1981.

Stafford, J., and Queen, C., Nature *306:*77, 1983.

SECTION **18-3.**

Electroporation

DESCRIPTION

This section describes a method for stably introducing a linearized plasmid into the genome of cultured mammalian cells. A high-voltage pulse is generated in a foil-lined cuvette containing the mammalian cells and linearized DNA. The cells become permeable to DNA and take up the exogenous DNA. Some of this absorbed DNA is integrated into the genome, creating a stable transformant. If the introduced plasmid contains a selectable marker, this stable integration event can be identified by growth in a selection medium. This method is used primarily for generating stable transformants. Transformation efficiencies vary between 10^{-4} and 10^{-6} transformants per cell, depending on the cell line used.

TIME REQUIRED

1 hr

SPECIAL EQUIPMENT

Electroporation apparatus (Shock Chamber)[1]

ISCO power supply model 494[2] or electroporation device from Prototype Design Services

---

[1] Shock chamber, as described, is homemade. Aluminum foil electrodes, 1 mil thick, are inserted in the opposite ends of a plastic cuvette. A discharge switch, suitable for high voltage and properly shielded, is placed in line with the circuit. Presterilize the chamber before use with 70% ethanol. A similar device is now commercially available from Bio-rad. Another apparatus is being marketed by Prototype Design Services, P.O. Box 55355, Madison, WI 53705 and has been reported to reproduce the results presented by Smithies et al. (see ''References'').

[2] The settings used in this method refer to the ISCO power supply. It may be possible to use other power supplies, depending on capacitor characteristics. One such substitution is described by Smithies et al.

REAGENTS

Phosphate-buffered saline (PBS)
Cell growth medium, depending on cell line

METHODS

In Advance

Linearize plasmid DNA by RE digestion (Section 5-4). Be sure not to cleave plasmid construct in the gene of interest or the selectable marker.

Cell Preparation

1. Harvest mammalian cells growing at mid-to-late log phase. Use 5 million cells per transfection.[3]
2. Concentrate cells by centrifugation at 700 × g for 5 min.
3. Remove medium. Resuspend pelleted cells in 5 ml of PBS at 4°C.
4. Repeat centrifugation. Resuspend cells in ice cold PBS at 10^7 cells per milliliter of PBS. Transfer cells to shock chamber on ice.
5. Add linearized DNA (1–10 μg), extracted and precipitated as in Section 20-1, in as small a volume as possible (under 5 μl).[4] Allow to sit for 5 min.
6. Connect power supply to shock chamber.[5,6] Set to 2,000 V and 0.9 mA limit. Set wattage and current dials to 5%.
7. Engage safety switch and discharge power supply across the shock chamber. When the power supply has been discharged, the voltage meter will read zero. Do not touch leads until voltage has reached zero.
8. Allow system to sit for 10 min on ice. Transfer cells, using sterile conditions, to a cell culture plate or flask and add 10 ml of growth medium at 4°C.
9. Grow cells for 48 hr without selection (use the proper growth medium and conditions for the cells used).
10. Transformants (cells with DNA incorporated) are selected in medium supplemented with appropriate selectable marker at a level sufficient to kill nontransformed cells. For example, if the neomycin gene was incorporated with the linearized plasmid, cells are selected with G418 (Section 18-4).

---

[3] Use only sterile or autoclaved tubes and buffers for this procedure. Wear gloves to avoid contamination. This method has been primarily used for B-lymphocytes, but other cells have also been used successfully. The parameters of the electric shock pulse may have to be altered with different cell lines.

[4] DNA should be at 2μg/μl. Generally, this is obtained from a pBR plasmid construct that also has a mammalian selectable marker gene (e.g., neomycin; Section 18-4).

[5] Handle the power supply and shock chamber with extreme caution during charging and discharge.

[6] See notes 1 and 2.

REFERENCES*

Falkner, F., and Zachau, H., Nature *310*:71, 1984.

Neumann, E., Gerisch, G., and Opatz, K., Naturwissenschaften *67*:414, 1980.

Neumann, E., Schaefer-Ridder, M., Wang, Y., and Hofschneider, P., EMBO J. *1*:841, 1982.

Potter, H., Weir, L., and Leder, P., Proc. Natl. Acad. Sci., USA *81*:7161, 1984.

Smithies, O., Gregg, R. G., Boggs, S. S., Koralewski, M. A., and Kucherlapati, R. S., Nature *317*:230, 1985.

---

* Dr. Hunt Potter provided technical information used in this method.

SECTION **18-4.**

Selection of Transfected Mammalian Cells: The G418 Method

DESCRIPTION AND GENERAL NOTES

This method uses a dominant selectable marker for the stable integration of transformed exogenous DNA, the 3′ phosphotransferase gene that confers G418 resistance. When cloned DNA is introduced into eukaryotic cells by transfection techniques such as $CaPO_4$, DEAE dextran, or electroporation (Sections 18-1, 18-2, and 18-3), only a small fraction of cells stably integrate the exogenous DNA into their genome. To identify these stable integration events, it is useful to have a dominant genetic marker in the transfected exogenous DNA that produces a selectable change in the phenotype of these cells. A number of selectable markers have been characterized and used successfully for this purpose. These include the thymidine kinase gene in TK^- cells and the dihydrofolate reductase gene, which confers resistance to the antibiotic methotrexate. Also used is the xanthine-guanine phosphoribosyl transferase gene, which permits xanthine to be used as a substrate in the purine salvage pathway and thus functions as a dominant selectable marker in an appropriate medium containing mycophenolic acid.

This section describes the properties and use of another dominant selectable marker for DNA transformation in mammalian cells. The mechanism of selection is based upon the cell's sensitivity to the aminoglycoside antibiotic G418. This antibiotic affects the function of 80S ribosomes and protein synthesis. G418 can be inactivated by the bacterial phosphotransferase enzyme APH(3′)II (neo), which is encoded on the transposon Tn5. When the coding sequences for this gene are arranged in a mammalian transcription unit to allow efficient expression, this chimeric gene confers resistance to G418 when it is stably integrated into the genome after transfection. If this marker gene (neo) is linked to or cointegrated with another gene whose stable integration is desired in the transformed cell, selection for resistance to G418 also selects for stable integration and expression of the desired gene.

Several features of this selection system are important for its effective use. Mammalian cells show a wide variation in their sensitivity to concentrations of G418 in the surrounding medium, ranging from about 100 μg/ml to 1 mg/ml. The amount of G418 needed for selection must be individually determined for each cell line. One simple method for determining an appropriate level is to grow the cells in a multiwell plate with a range of G418 concentrations in the medium. The appropriate level of G418 for selection is the lowest concentration that kills the mammalian cells within 10–14 days. Cells that divide more rapidly are typically killed more readily in G418 selection because the aminoglycoside appears to act primarily on dividing cells.

One very convenient source of a neomycin gene that is expressed in most mammalian host cells is the plasmid pSV2-neo, constructed by Southern and Berg (see "References").[1] In this plasmid, the 1.4-kb neomycin coding sequence from Tn5 is sandwiched between the simian virus 40 (SV40) early region promoter and the SV40 small t antigen splice and polyadenylation site. A chimeric mRNA (about 2 kb) is transcribed from the SV40 promoter and stabilized by the SV40 splice and polyadenylation signals located 3′ to the neomycin coding region. Stable integration of a cloned gene whose expression is desired in a mammalian cell host can be achieved by covalently attaching the relevant DNA to the pSV2-neo plasmid at a variety of available sites. Alternatively, the gene to be studied can be cotransformed as a separate plasmid with the pSV2-neo plasmid. Stable integration of both plasmids will frequently result, allowing selection in media containing G418. Transformation efficiencies for CV-1 and HeLa cells using $CaPO_4$ precipitation (10 μg of DNA in 5×10^6 cells) are generally on the order of one transformant per 10^4 to 10^5 cells. Efficiency will vary depending on the eukaryotic host cell; different transformation techniques may give higher or lower efficiencies with different cell lines, and the optimal system may have to be determined empirically. The spontaneous frequency of G418 resistance in mammalian cells is reported to be below 10^{-7}, making this selection system practical for many cells (see Southern and Berg in References below).

In addition, the 3′ phosphotransferase gene can be used to select for G418 resistance in *E. coli.* The levels of G418 or kanamycin required for selection are usually lower than those in eukaryotic cells, in the range of 5–20 μg/ml. This selection sometimes proves useful when constructing plasmids containing the neo gene for subsequent eukaryotic cell transformation.

REFERENCES

Collbere-Gorapin, F., Horodniceanu, F. Kowilsky, P., and Garapin, A., J. Mol. Biol. *150:*1, 1981.

Jimenez, A., and Davies, J., Nature *287:*869, 1980.

Southern, P. J., and Berg, P., J. Mol. Appl. Gen. *1:*327, 1982.

---

[1] A plasmid construction containing the neo selection gene is available from Pharmacia, Inc.

SECTION **18-5.**

Chloramphenicol Acetyltransferase (CAT) Assay*

DESCRIPTION

This method is used primarily for the identification of potential eukaryotic gene regulatory elements, such as promoters and enhancers. The elements to be tested are fused to the coding region of the bacterial CAT gene present on a recombinant DNA plasmid. This plasmid is transfected into mammalian cells by the $CaPO_4$ or DEAE dextran method, and the level of CAT enzyme activity is subsequently assayed by autoradiographic determination of radiolabeled enzyme products. Thus, an active promoter or enhancer on the cloned DNA of interest will also activate CAT activity when transfected into an appropriate eukaryotic host.

TIME REQUIRED

Day 1—2 hr
Day 2—4 hr

SPECIAL EQUIPMENT

Autoradiography equipment
Sonicator for cell disruption
Chromatography tank
Rotary evaporator (e.g., SpeedVac Concentrator by Savant)
Liquid scintillation counter

---

* This method was contributed by Dr. Shoshana Segal.

REAGENTS

Phosphate-buffered saline (PBS)
0.25 M Tris, pH 7.8
^{14}C-Chloramphenicol, 40–60 mCi/mmole (NEN)
4 mM acetyl coenzyme A (Sigma) Make fresh every 2 weeks and store at −20°C.
Ethyl acetate[1]
Chloroform
Methanol
Thin layer chromatography (TLC) plates or sheets (e.g., Baker Flex TLC)

METHODS

In Advance

Prepare plasmid vector by fusing CAT-containing plasmid with DNA element to be tested.[2] Grow eukaryotic cells to be transfected.[3] Prepare TLC tank by placing a sheet of Whatman 3MM filter paper around the inside of the tank and add 200 ml of chloroform:methanol (19:1) solvent to equilibrate tank atmosphere. Make solvent system fresh for each use, because chloroform is very volatile.[4]

Transfection of Cells

1. Transfect eukaryotic cells (approximately 1×10^7 cells) with the CAT construct (25–50 μg of plasmid DNA) using the $CaPO_4$ or DEAE Dextran method (Sections 18-1 or 18-2).[5]
2. After 36–48 hr, harvest the cells and wash three times with PBS.
3. Resuspend cells in 100 μl of 0.25 M Tris, pH 7.8.
4. Break open cells by sonication. Check cells under microscope to make sure they have been broken.[6]

---

1 Ethyl acetate dissolves many plastics. Use glass pipettes with ethyl acetate solutions. Use ethyl acetate in a fume hood.

2 The sequences being examined for promoter/enhancer activity are typically ligated to the CAT coding region that is located on an appropriate plasmid vector (such as pSVO-CAT or pSV2-CAT; see the reference); pBR327 plasmids containing a derivative of the CAT gene are available from Pharmacia, Inc.

3 The assay described here has been standardized for extracts from CV-1, 3T3, HeLa, and Chinese hamster ovary (CHO) cells (approximately 10^7 cells). The procedure can be scaled up or down, depending on the number of cells to be analyzed.

4 Perform chromatography in a fume hood.

5 For the CAT assay, $CaPO_4$ transfection usually works better than the DEAE dextran method. Some cell lines appear to be more efficiently transfected with one of the two methods.

6 A small sonicator probe is used to break open cells. Yield of broken cells can be determined by trypan blue exclusion.

5. Spin for 5 min in a microcentrifuge at 4°C to remove cellular debris.
6. Transfer supernatant to a new microfuge tube. Incubate for 10 min at 60°C to inactivate endogenous acetylases.
7. Spin for 5 min in a microcentrifuge at 4°C. Transfer clear supernatant to a new microfuge tube. Supernatant is now ready for CAT assay.

CAT Assay

8. Combine the following reagents in a new microfuge tube:

   25 μl of cell extract from step 7
   5 μl of 4 mM acetyl coenzyme A
   18 μl of H_2O
   1 μl of ^{14}C-chloramphenicol

9. Incubate for 1 hr at 37°C.
10. Add 400 μl of ethyl acetate. Mix well by vortexing. Spin for 1 min in a microcentrifuge. Remove upper organic phase to a fresh tube.
11. Dry sample under vacuum with rotary evaporator.
12. Resuspend sample in 25 μl of ethyl acetate.
13. Spot sample on a TLC plate (or sheet). Also spot 1 μl of ^{14}C-chloramphenicol as a reference.
14. Place TLC plate or sheet in a chromatography tank that has been preequilibrated with 200 ml of chloroform:methanol (19:1)—see "In Advance"—and chromatograph until the solvent front is two-thirds of the way up the plate or sheet.
15. Air dry and expose to X-ray film.
16. Develop film. The presence of active CAT activity will yield a second acetyl chloramphenicol spot running further than the ^{14}C-chloramphenicol spot. The relative intensity of the acetyl-chloramphenicol compared to the chloramphenicol spots will correspond to the CAT activity. Spots corresponding to acetylated and unacetylated ^{14}C-chloramphenicol may be excised from the TLC plate or sheet and counted by liquid scintillation for quantitation of CAT activity.

REFERENCE

Gorman, C. M., Moffat, L. F., and Howard, B. H., Mol. Cell. Biol. *2*:1044, 1982.

SECTION

19

Protein Methods

SECTION

19-1. In Vitro Translation and Immunoprecipitation

DESCRIPTION

This method allows the investigator to identify potential translation products from partially purified mRNA. It translates RNA into protein that is subsequently analyzed by precipitation with an antibody and polyacrylamide gel electrophoresis.

TIME REQUIRED

Day 1— 2 hr
Day 2—5 hr

SPECIAL EQUIPMENT

Scintillation counter
Vertical polyacrylamide gel apparatus

REAGENTS

Reticulocyte mix (Amersham Kit N.90)[1]
^{35}S-methionine, high specific activity, 800 Ci/mmol (NEN, Amersham)
10% trichloracetic acid (TCA)
Ethanol
Protein A-Sepharose (P-L)
Normal rabbit serum

---

[1] Other translation methods exist, such as the wheat germ translation kit. This alternative is very useful for small protein products (up to approximately 20,000 daltons).

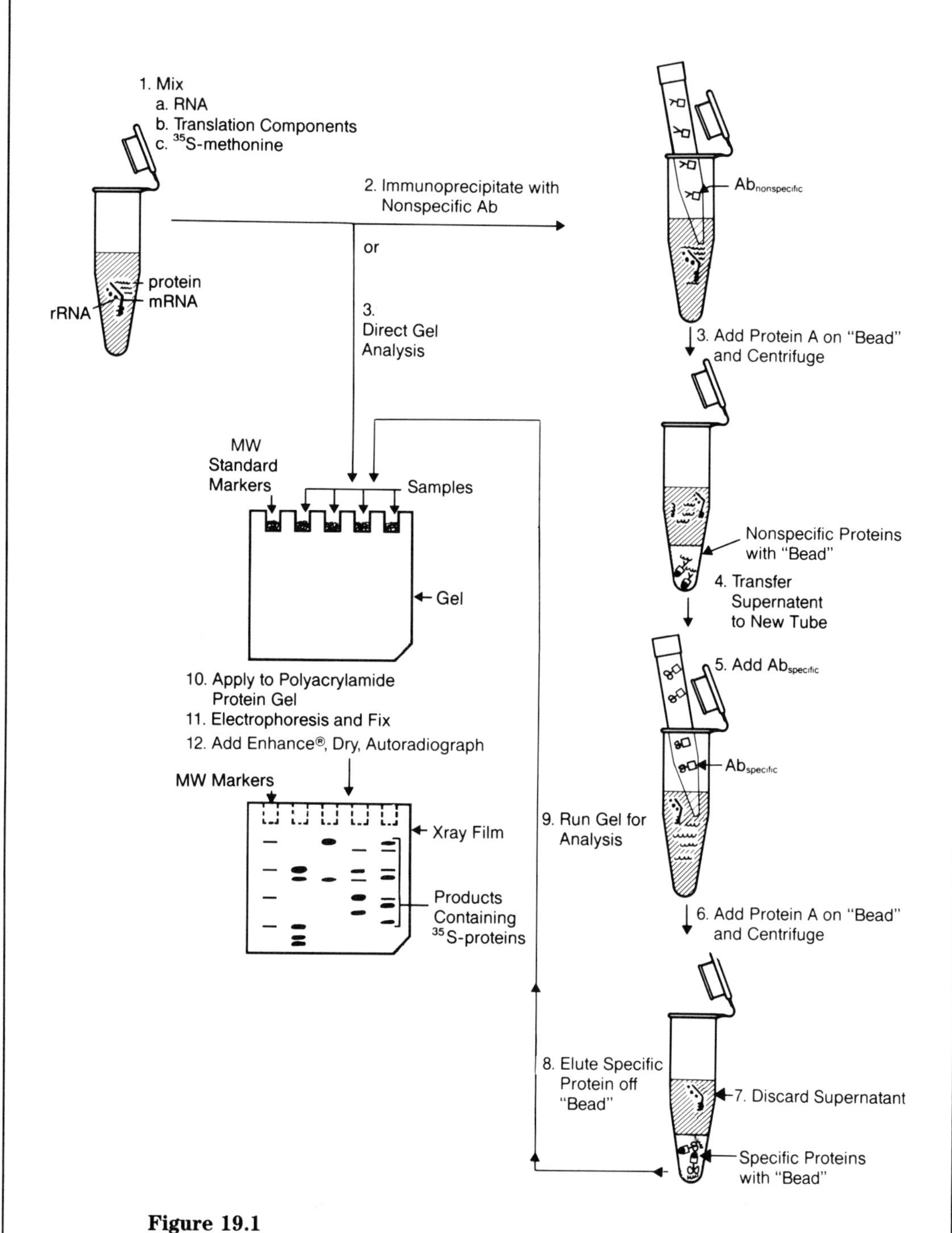

Figure 19.1
In vitro translation of RNA into radiolabelled protein. The products can be selected by immunoprecipitation (right half) or directly analyzed on a polyacrylamide gel for size analysis. ("Bead" is the inert support, such as Sepharose beads.)

Immunoprecipitation buffer

10 mM Tris, pH 7.4

0.5% aprotinin

1% Non-idet P-40 (Sigma)

2 mM EDTA

0.15 M NaCl

Antiserum for desired protein product

Sample buffer is polyacrylamide sample buffer from Section 19-2, diluted to 1×

Enhance or Enlightening (NEN)

METHODS

1. Heat total RNA sample (10 μg in 2 μl) from Section 11-1 or poly(A^+) RNA from Section 11-3 (2 μg in 2 μl) in a 65°C water bath for 2 min in a 1.5-ml microfuge tube to denature RNA sample.
2. As described in reticulocyte kit instructions, include a blank (no RNA) and a positive control (e.g., globin RNA) tube along with sample tubes.
3. Add 20 μl of reticulocyte mix from kit and 5 μl of ^{35}S-methionine. Briefly spin in microcentrifuge.
4. Incubate for 1 hr at 35°C.[2]
5. To determine effectiveness of translation, spot 2 μl of sample on small paper filter. Also spot blank and control samples on additional filter papers. Freeze remaining sample at −20°C.
6. Soak filter paper in 10% TCA, prewarmed to 65°C for 5 min. Repeat soaking in fresh warmed 10% TCA for 5 min.
7. Wash filter paper in absolute ethanol and count filter paper in scintillation counter. If counts from sample are at least three times the counts of the blank, translation has occurred. If so, proceed to step 8. If not, determine whether the positive control worked (it should be approximately seven times the blank). If the positive control in the kit incorporated radioactivity and the RNA sample did not, then the RNA sample was probably inadequate. If the positive sample didn't incorporate radioactivity, the kit may be inactive.
8. The protein translation product can now be employed in one of two ways. It can be immunoprecipitated with a known antiserum for identification (proceed to step 9), or the total translated products can be analyzed on polyacrylamide gels (Section 19-2) for direct analysis by autoradiography.

---

[2] One hour is a typical time. The optimal time for incubation may vary, depending on RNA quality and translation materials.

Precipitation and Gel Separation

9. Thaw the remaining translation products, add 600 μl of immunoprecipitation buffer. Divide into two aliquots (300 μl each) in 1.5-ml microfuge tubes.
10. Add 2 μl of normal rabbit serum[3] to each tube and incubate for 10 min at room temperature.
11. Add 40 μl protein A-Sepharose to each tube and rock gently at 4°C for 30 min.
12. Spin in microcentrifuge for 1 min.
13. Transfer supernatant to new tubes. Discard old tubes with pellet.
14. To one tube, add 2 μl normal rabbit serum. To the other tube, add 2 μl of the desired antiserum at an appropriate dilution.[4] Incubate overnight at 4°C.
15. To both tubes, add 80 μl of protein A-Sepharose. Incubate at 4°C for 30 min with gentle rocking. *Rocking is important here.*
16. Spin in microcentrifuge for 1 min. Discard supernatant.
17. Wash each pellet with 300 μl of immunoprecipitation buffer. Shake gently. Spin in microcentrifuge for 1 min. Discard supernatant. Repeat this step three more times.
18. To each pellet add 40 μl of sample buffer and place in boiling water bath for 2 min.
19. Spin in microcentrifuge for 5 min. Load supernatant on polyacrylamide gel and run gel (see Section 19-2).
20. For efficient autoradiography, incubate gel for 30 min in Enhance, as per manufacturer's instructions.
21. Dry gel. Expose film for autoradiography.
22. Develop film. Adjust exposure time and reexpose, if necessary.

REFERENCE

Persson, H., Hennighausen, L., Taub, R., DeGrado, W., and Leder, P., Science *225*:287, 1984.

---

[3] This typically comes from prebleeding the same rabbit before immunization, but normal rabbit serum can be purchased.

[4] The dilution is empirically determined. This is the largest dilution that will effectively precipitate the protein sample. If titer information is not available, use multiple dilutions to determine best concentration for further use.

SECTION **19-2.**

Polyacrylamide Gels for Protein Separation

DESCRIPTION

It is often useful to resolve individual proteins from a heterogenous mixture based on their molecular weights in order to visualize their overall composition or to evaluate protein purity from an immunoprecipitation. This method describes the separation of proteins in an electric field by inducing their movement through an acrylamide matrix. The effect of different protein charges is minimized by their uniform association with SDS, and thus the separation is due principally to differences in protein size (molecular weight).

TIME REQUIRED

1 day

SPECIAL EQUIPMENT

Protein gel electrophoresis apparatus, plates, spacers, and so on (available from many suppliers)

Gel dryer

REAGENTS

Acrylamide:bisacrylamide (29:1) prepared as an unpolymerized 30% solution (wt/vol) in H_2O. Handle with extreme caution, acrylamide is a neurotoxin.

Mixed ionic resin: AG501-X8 (Biorad)

Molecular weight standards (high and low set) (e.g., Biorad)

Ammonium persulfate, 100 mg/ml, made 1–2 days before use

TEMED

10% glycerol, 300 ml

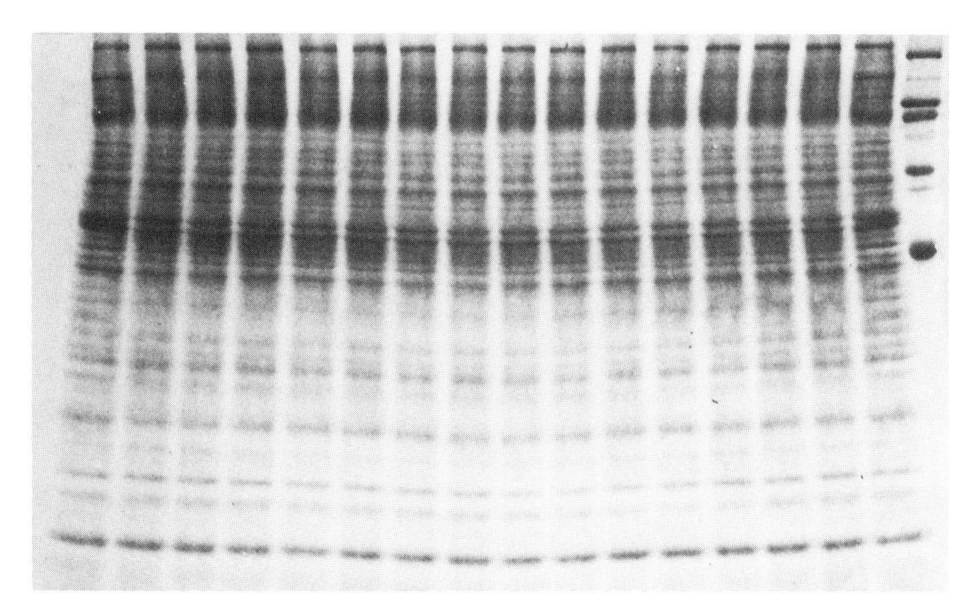

Figure 19.2
Coomassie blue staining of protein in a polyacrylamide gel. Multiple lanes of the same sample. Note size markers in right lane.

4× gel buffer

16.95 g of Tris base (1.5 M)
Add H_2O to 90 ml
pH to 8.8 (with HCl)
4 ml of *10% SDS* (0.4%)
Bring up to 100 ml with H_2O.

10× electrode buffer

30.25 g Tris base (0.25 M)
144 g glycine
Add H_2O to 850 ml
Check pH, should be 8.3
100 ml of *10% SDS*
Bring up to 1 liter with H_2O.

2× stacking gel buffer

3 g Tris base (0.25 M)
Add H_2O to 90 ml
pH to 6.8 (with HCl)
2 ml *10% SDS (0.2%)*
Bring volume to 100 ml with H_2O.

2× sample buffer
- 4 ml H_2O
- 0.15 g Tris base (0.125 M)
- pH to 6.8 with HCl
- 4 ml *10% SDS (4%)*
- 1 ml glycerol (10%)
- 20 mg bromophenol blue (0.02%)
- 0.4 ml β-mercaptoethanol (4%)
- Bring volume to 10 ml with H_2O.

Wash solution for gels (600 ml needed per gel)
- 50% methanol
- 10% acetic acid
- 40% H_2O

Coomassie brilliant blue R (Sigma)

METHODS

In Advance

Deionize 1 liter of 30% acrylamide:bisacrylamide (29:1) by adding 20 g mixed ionic resin, gentle stirring, and filtering through Whatman paper filter or 0.22-μm filtration unit to remove resin. Stored at 4°C in a foil-wrapped bottle.

Prepare plates for protein gel. Wash sequentially in H_2O, 10% SDS, H_2O, ethanol, and H_2O. Air-dry the cleaned plates.

Prepare Gel

1. Mix 25 ml of deionized 30% acrylamide/bisacrylamide with 15 ml of 4× running gel buffer and 100 μl of ammonium persulfate. Bring volume to 60 ml with H_2O. This gives a 12.5% gel; the percentage of acrylamide can be changed, depending on the protein size.[1]

---

[1] A concentration of 12.5% is a suitable starting point because it is effective for proteins in the 50,000-dalton range. Sixteen percent (32 ml of 30% acrylamide) is good for 10,000–30,000 daltons, and 8% (16 ml of 30% acrylamide/bisacrylamide) can be used for 80,000 daltons. Other components and the final volume remain the same. For optimal separation of total proteins over a wide range of sizes, two solutions of 8% and 16% acrylamide can be used to make a gradient gel with a gradient former by linearly mixing the 8% into the 16% acrylamide and filling the gel over 20 min from the bottom with a 3 ml/min flow rate through a capillary tube attached with tape to the gel plate. Note that only 30 ml of each (1/2 volume) needs to be made to fill the gel space properly, if a gradient is desired. Acrylamide is a neurotoxin; handle with care and wear gloves. Also, do not inhale acrylamide in powdered form.

2. Attach clean glass plates to each other, separated by lightly greased spacer. Different gel systems may have slight variations. Follow manufacturer's instructions.
3. Add 30 μl TEMED to catalyze polymerization. Pour the 60 ml into the prepared plates and gently layer a few mililiters of 2× stacking gel buffer on top to form a flat surface while acrylamide polymerizes.[2]
4. After 1 hr, pour the upper unpolymerized liquid out. Prepare a top "stacking" gel. Mix 1 ml of 30% acrylamide/bisacrylamide (29:1), 4 ml of H_2O, 5 ml of 2× stacking gel buffer, 10 μl ammonium persulfate, and 5 μl TEMED. Pour enough of this stacking gel mixture to fill the upper 3 cm of the gel. Insert sample position forming comb, and allow "stacking" gel to polymerize. The addition of the stacking gel sharpens the bands and increases resolution.
5. Attach gel to electrophoresis unit. Add approximately 500 ml 1× electrode buffer to both top and bottom chambers and check for leaks.[3]
6. Clean sample wells by pipetting electrode buffer in and out of each.
7. Mix 10 μl of sample (1–100 μg protein)[4] with 10 μl of 2× sample buffer. Heat to 95°C for 3 min and load samples on gel. Include molecular weight standards in one or two lanes.
8. Run gel until blue dye front just runs out of the gel. The typical setting is 100 constant V, which should run in about 4-6 hr. Time and voltage with depend on apparatus and gel thickness.
9. Gel can be stained with silver stain (Section 19-4) or with Coomassie brilliant blue (less sensitive, but easier).[5] If radioactive amino acids have been incorporated, process for autoradiography (steps 10–14, with or without staining).

Preparation of Gel for Autoradiography

10. Remove gel from plates. Mark for orientation by cutting off a small piece of the lower right corner. Place gel in container. Add 300 ml of wash solution. Let stand, with periodic rocking, for 1 hr.
11. Discard liquid and add 300 ml of fresh wash solution. Let stand at least 2 hr.

---

[2] Wear gloves for handling samples and gels to prevent contamination. If sample is radioactive, handle reagents, gel, and buffer accordingly.

[3] Final concentration of SDS in gel buffer is 0.1%.

[4] Sample can be protein from various sources in aqueous buffers.

[5] For Coomassie brilliant blue staining, follow steps 10 and 11, then add 300 ml of 0.2% Coomassie brilliant blue in 10% acetic acid, 50% methanol, and 40% H_2O. Incubate for 4 hr or more. Destain gel (remove background while still being able to visualize bands) by repeating steps 10 and 11 three times with intermittent mixing (a Hoefer charcoal destainer may be useful). The final wash step is in 10% acetic acid to reswell gel, followed by incubation in 10% glycerol before drying.

12. Discard wash solution.[6] Add 300 ml of 10% glycerol. Let stand for 1–2 hr.
13. Discard glycerol solution. Dry gel at 80°C under vacuum for 2 hr in gel dryer. Turn off heat and keep under vacuum for an additional 30 min.[7] Be careful not to disturb the vacuum until the gel is dry, or the gel may crack.
14. If samples are radioactive, expose to X-ray film (XAR-5) for 1–3 days. Use screen at −70°C for ^{32}P-labeled samples. Process film.[8]

REFERENCES

Laemmli, U.-K., Nature *227*:680, 1970.

Mahadik, S. P., Korenovsky, A., and Rapport, M. M., Anal. Biochem. *76*:615, 1976.

---

[6] If fluorography is desired to increase the sensitivity of autoradiography of ^{3}H, ^{14}C, or ^{35}S Enhance or Enlighten from NEN can be used. Follow instructions with product: shake gel in this solution for 30 min and then dry gel (step 12).

[7] Drying time may vary, depending on gel thickness and strength of vacuum pump.

[8] Exposure time depends on amount of labeled protein present in gel. See Section 20-5 for more details.

SECTION **19-3.**

Western Blot Analysis

DESCRIPTION

This method allows the investigator to identify specific proteins resolved by SDS polyacrylamide gel electrophoresis by binding with specific antisera. Proteins resolved on an acrylamide gel are transferred to an NC filter that is incubated with the antisera. The primary antibody specifically binds its epitope and the bound antibody is detected with a secondary species, such as [^{125}I]-protein A or biotinylated goat anti-IgG.

TIME REQUIRED

2 hr or overnight for electrophoretic transfer

7 hr for hybridization and visualization

SPECIAL EQUIPMENT

Electrophoretic transfer apparatus[1] (e.g., Biorad, Hoefer, E-C apparatus)

Plastic bag sealer

Rocking apparatus

REAGENTS

Amido black stain, 0.2% in 7% acetic acid

NGS (normal goat serum), heat inactivated at 56°C for 60 min, filtered through a 0.22-μm filter

---

1 Transfer electrophoresis setup and running conditions vary widely among manufacturers.

Transfer buffer

12 g Tris base

57.65 g glycine

4 liters H_2O

After Tris and glycine are dissolved, add 1 liter methanol.

pH to 8.3 with HCl

Destaining buffer (optional)

30% methanol and 10% acetic acid in H_2O

Blot buffer

5% (wt/vol) Carnation nonfat dry milk in PBS (also called BLOTTO buffer)

Phosphate-buffered saline (PBS)

PBS-Tween buffer

0.05% Tween-20 in PBS

(2.5 ml 20% Tween-20 in 1 liter PBS)

Primary antibody: antisera made against protein(s) of interest, this may be monoclonal or heterologous antisera.

Secondary antibody

Biotinylated goat anti-primary IgG (Vector Labs)[2]

Avidin-horseradish peroxidase (HRP) (Vector Labs)

Substrate

30% hydrogen peroxide diluted to 1% (1:30) in H_2O

Color indicator stock solution

4-chloro-1-napthol (3 mg/ml H_2O)

METHODS

In Advance

Prepare polyacrylamide-SDS gel for separation of proteins (see Section 19-2). Obtain antibody against protein sequence of interest.

---

[2] There are numerous ways to detect specific protein(s) with antisera. The example described here is as follows:

Rabbit Ab + goat antirabbit-biotin + avidin-HRP + color-peroxide.

This can be substituted for by antibodies from other species with a suitable biotinylated second antibody. Alternatively, other indicator systems can be attached to the primary or secondary antibody.

Western Blot

1. Electrophoretically transfer protein in polyacrylamide gel to NC filter:

   a. Rinse both the gel and the NC filter in transfer buffer. Place NC filter against gel on flat surface. Smooth filter over gel by rolling with a 5-ml glass pipette to remove all air bubbles.
   b. Wrap a piece of 3MM filter paper (prewetted with transfer buffer) around gel/filter to make a sandwich. Keep wet and avoid bubbles.
   c. Paper/gel/filter/paper sandwich is placed in electrophoretic transfer apparatus, as suggested by the manufacturer, with the gel facing the cathode.
   d. Place in buffer tank and fill with transfer buffer to cover gel.
   e. Connect to power supply and run electrophoretic transfer, as suggested by manufacturer.[3]
   f. After transfer is complete, remove filter and gel from unit. Discard gel.

2. Float filter in amido black stain briefly until molecular weight standards become visible. Mark standard positions.
3. Wash NC briefly in 100 ml H_2O. Use destaining buffer if necessary.
4. Incubate filter in blot buffer for 1 hr at 37°C.
5. Wash in PBS-Tween buffer for 1 hr at room temperature.
6. Seal filter inside plastic bag with heat seal apparatus. Seal as close as possible to filter to keep the inside volume small.
7. Cut corner of bag to use as buffer entry port. Close hole with dialysis clamp.

Antibody Reaction

8. Mix:

   100 μl NGS
   Antibody (dilution will vary with antibody)[4,5,6] in 10 ml of blot buffer

   Add to bag with filter. Incubate for 2 hr at room temperature (or overnight at 4°C) with rocking.[7]
9. Wash filter in a shallow dish with four changes of 75 ml of PBS-Tween buffer for a total of 300 ml over 30 min.

---

3 Typical settings are 0.1 Amp for 16 hr or 0.7 Amp for 1 hr.

4 See note 2.

5 Try multiple dilutions to determine empirically the optimal antibody concentration.

6 Control serum or preimmune serum can be used as a control.

7 Rock during all steps for efficient mixing (in rocking apparatus).

10. Add biotinylated goat anti-rabbit IgG[8,9,10] (40 μl in 10-ml blot buffer with 100 μl NGS) to bag. Incubate for 1 hr at room temperature with rocking.
11. Wash, as in step 9.[11]
12. Add avidin-HRP (40 μl in 10 ml of blot buffer with 100 μl of NGS). Incubate for 30 min at room temperature with rocking.
13. Wash, as in step 9.
14. Make color indicator/substrate

    Mix: 3 ml color indicator stock solution
    9 ml PBS
    Add: 150 μl of 1% hydrogen peroxide

    Use immediately.
15. Incubate at room temperature with rocking until purple color develops and background begins to be detectable (1–10 min).
16. Transfer filter and wash in another container containing H_2O to halt color development. Air-dry. Store in a plastic bag in the dark.

REFERENCES*

Johnson, D. A., Gautsch, J. W., Sportsman, J. R., and Elder, J. H. Gene Anal. Tech. *1*:3, 1984.

Towbin, H. H., Staehelin, T., and Gordon, J. Proc. Natl. Acad. Sci, USA *76*:4350, 1979.

---

[8] See note 2.

[9] See note 5.

[10] Instead of biotinylated second antibody, a ^{125}I-labeled second antibody or protein A can be used.

[11] If ^{125}I label is used, expose gel to X-ray film after drying. It is not necessary to proceed to step 12.

* D. Reed provided technical information for preparing these methods.

SECTION **19-4.**

Silver Staining of Gels for Proteins or RNA

DESCRIPTION

This section describes a sensitive and efficient way to visualize RNA or protein bands on polyacrylamide gels. Two methods are presented: the use of a kit (e.g., Biorad or NEN) and a less expensive laboratory method.

TIME REQUIRED

2–3.5 hr

REAGENTS

Kit Method

Silver Stain Kit (Biorad or NEN) with oxidizer, silver stain reagent, and developer
40% methanol
10% ethanol
5 and 10% acetic acid

Nonkit Method

Fixative 1—50% methanol with 7.5% acetic acid
Fixative 2—5% methanol with 7.5% acetic acid
Fixative 3—10% glutaraldehyde
Developer
  0.05% citric acid
  Add 5 μl of a 37% formaldehyde solution for each ml of 0.05% citric acid

Stain: To make 250 ml:

53 ml of 0.09 M NaOH

3.5 ml of NH_4OH (concentrated)

Very slowly add: 8 ml of $AgNO_3$ solution (20 g per 100 ml H_2O); swirl while adding

Bring volume to 250 ml with H_2O

METHODS

In Advance

For RNA, run polyacrylamide gel (1.0-mm thickness) on electrophoresis apparatus.[1] For protein, run 1.0- to 1.5-mm gel, as indicated in Section 19-2. Run 10–100 ng per well.

Kit Method (all steps are performed at room temperature with shaking)[2]

1. After running electrophoresis, place gel in plastic dish with 400 ml of 40% methanol. Change reagents in dish at the indicated times, volumes, and solutions:[3]

Solution	Volume (ml)	Protein Time (min)	RNA Time (min)
40% methanol/10% acetic acid*	400	60	30
10% ethanol/5% acetic acid*	400	30	15
10% ethanol/5% acetic acid*	400	30	15
Oxidizer (from kit)	200	10	5
H_2O	400	10	10
Repeat H_2O		2 times	1 time
Silver stain reagent (from kit)	200	30	20
H_2O	400	2	10
Developer (from kit)	200	1	0.5
Developer	200	5	5
Developer	200	5	5
5% acetic acid (stop solution)	400	5	5

* Do not use acetic acid in the first three steps for RNA staining.

[1] RNA gel is assembled as in Section 19-2, except that SDS is not used in gel and 1× TBE is used as a running buffer. Gel is 5% acrylamide and 8 M urea in 1× TBE, as in Section 16-9. Use gel loading solutions as in Section 16-9. Run at 400 V.

[2] Follow recommendations and steps indicated with kit. Times and volumes shown here are typical.

[3] Do not touch gel with hands, except on edges, even if gloves are worn; finger marks will be left. A fly swatter is useful to pick up gel when changing solution.

2. Bands will develop during the last three steps. Time of band development can vary; add stop solution before background becomes darker than desired bands. After stopping, wash gel in 400 ml of H_2O.[4,5] Gel can now be photographed or used for other analysis. Gel can be stored in sealed plastic bag.

Laboratory Nonkit Method for Staining Proteins

1. Stain gel in the following solutions:

Solution	Volume (ml)	0.75 mm Gel Thickness	1.5 mm Gel Thickness
		Time	
Fixative 1	400	20	30
Fixative 2	400	20	30
Fixative 3	400	20	30
Rinse in large volumes of H_2O, two to three times, preferably overnight in 2–3 liters of H_2O. This is very important.			
Stain	200	20	10
Watch carefully; if gel starts to turn brown, go immediately to next step.			
H_2O wash	500	5	5
Developer	250	5	5
(5 min is approximate. Develop until bands become visible; for slower development, bring volume to 400 ml with H_2O.)			
Wash in H_2O	400	5	5
Repeat previous step.			
Wash in large volume (e.g., 2 liters) of H_2O and change H_2O twice over 2 hr.*			
Photograph and store gel in a sealed plastic bag, if desired.			

* See notes 4 and 5.

REFERENCES*

Oakley, B. R., Kirsch, D. R., and Morris, N. R., Anal. Biochem. *105:*361, 1980.

Merril, C. R., Goldman, D., and Van Keuren, M. L., Meth. Enzymol. *96:*230, 1983

[4] Another batch of developer is used if further development is necessary.

[5] If overstaining occurs, use reducer to destain (from Kodak, Biorad, Farmers, etc.). If understaining occurs, try to stain again; rinse with H_2O and start at staining step in procedure.

* D. Reed provided technical information for the nonkit method.

SECTION

20

General Methods

SECTION

20-1. DNA/RNA Extraction and Precipitation

DESCRIPTION

To remove protein and contaminants from samples of DNA or RNA prior to use, an SS-phenol/chloroform extraction of protein is followed by an ethanol precipitation step. This method is a part of many methods found throughout this book.

TIME REQUIRED

Approximately 1 hr.

REAGENTS

SS phenol
Chloroform
Absolute and 80% ethanol
3 M sodium acetate, pH 7 for DNA
3 M sodium acetate, pH 6 for RNA
tRNA, yeast, 10 mg/ml

METHODS

In Advance

Prepare DNA or RNA to be extracted. The preparation is often in small volumes in a 1.5-ml microfuge tube, but can be in larger volumes in larger tubes. Preparation must be in an aqueous solution of known volume.

SS-Phenol/Chloroform

1. Add to each sample 1 volume of SS-phenol plus 1 volume of chloroform. Mix by vortexing.[1]
2. Centrifuge each tube for 1 min.[2]
3. Remove upper aqueous layer and transfer to new tube. For best results, angle tube to 45° before pipetting. Discard lower organic layer and interface.[3,4]

Ethanol Precipitation of DNA or RNA

4. To aqueous layer from step 3, add 1/10th of 1 volume of 3 M sodium acetate—pH 7 for DNA or pH 6 for RNA. This gives a final concentration of 0.3 M sodium acetate.[5,6]

---

[1] Mix gently for large fragments of DNA to avoid shearing. Mix these samples by repeated gentle inversions. Do not vortex.

[2] Microcentrifuge kept in a cold room at 4°C is preferred. For larger volumes, spin in larger tubes at 10,000 × g for 5 min at 4°C.

[3] SS-phenol is salt and water saturated and will become the bottom layer in an aqueous extraction. (Note that SS-phenol is the top layer in a stock bottle.) It is also colored yellow for easy identification. Some procedures call for the addition of small amounts of isoamyl alcohol in this step. A small amount of aqueous layer may be lost with each extraction. Adjust volumes accordingly. Some procedures call for a first extraction with 1 volume of a 1:1 mixture of chloroform:SS-phenol followed by a second chloroform extraction.

[4] Larger samples can be transferred in aliquots of 450 μl or less to 1.5-ml microfuge tubes. Alternatively, ethanol precipitation of larger samples can be done in larger (e.g., 10- to 15-ml) capped tubes.

[5] Instead of sodium acetate, 1/10th of 1 volume of 5 M NaCl can be used. Salt is necessary to form a nucleic acid precipitate at low nucleic acid concentrations. Sodium acetate may be preferred over NaCl for its buffering capacity.

[6] If the DNA or RNA concentration is less than 1 μg/ml, addition of yeast tRNA carrier (e.g., 10 μg in 1 μl) will increase recovery by facilitating precipitation. Be careful not to use tRNA carrier if it will interfere with subsequent procedures.

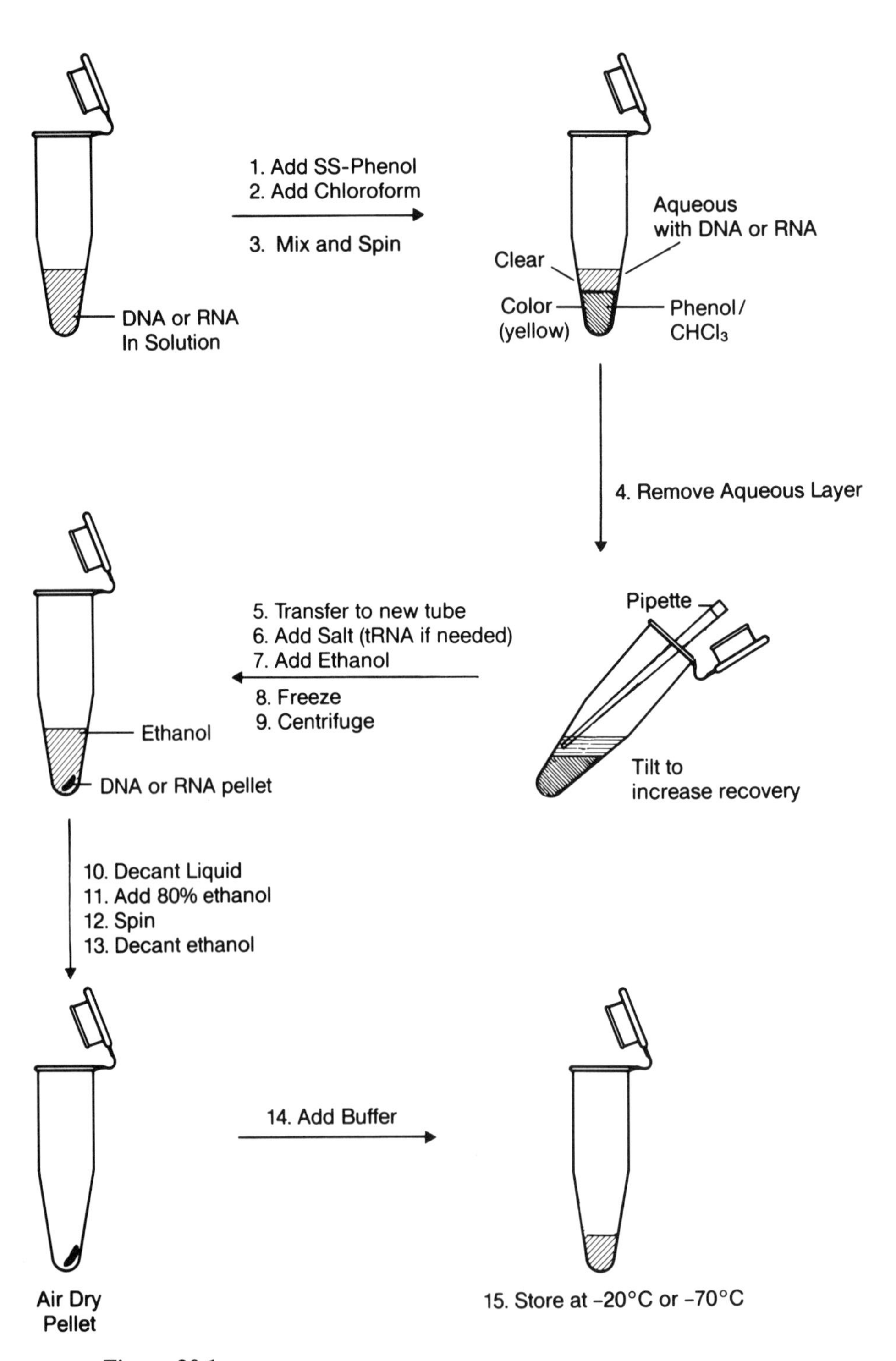

Figure 20.1
Purification of DNA or RNA. The sample is extracted with SS-phenol and chloroform to remove impurities such as protein, and then the nucleic acid is precipitated with ethanol.

5. Add at least 2.5 volumes of 95% ethanol to tube. Place on dry ice for at least 10 min or store for several hours at −20°C. Spin in microcentrifuge for 10 min at 4°C.[7]
6. Discard supernatant carefully; do not disturb the pellet. To remove residual salt, wash pellet in 300 μl or more of 80% ethanol. Mix by vortexing.
7. Spin in microcentrifuge for 2 min at 4°C.
8. Discard supernatant. Let pellet air-dry.[8]
9. Resuspend as indicated.[9]

REFERENCE

Kirby, K. S., Biochem. J. *64*:405, 1956.

---

[7] Frozen ethanol mix will thaw during centrifugation. If nucleic acid concentrations are sufficiently high in the initial aqueous solution (i.e., over 10 μg/ml), the nucleic acid precipitate will form with an incubation on ice (0–4°C) for 10 min. If concentration is lower, dry ice incubation is required.

[8] Large DNA pieces (e.g., genomic DNA or bacteriophage λ) should not be dried completely. Other samples can be dried completely using a vacuum evaporator.

[9] Suspension of DNA is often in TE buffer. It can be stored as a pellet stably at −20°C. RNA is often stored in H_2O or in 80% ethanol for increased stability. Store RNA at −70°C.

SECTION
20-2. Plastic Bag Sealing

DESCRIPTION

This is a technique for quick, easy, and efficient sealing of filters in bags for hybridization.[1]

TIME REQUIRED

5 min

SPECIAL EQUIPMENT

Plastic bag sealer (e.g., Sears, Krups)
Heat-sealable bags (e.g., Kapak)

METHODS

In Advance

Prepare blotted NC filters and hybridization buffer[2]

Procedure

1. Place filter(s) flat in a heat-sealable polyethylene bag.[3,4] Add hybridization buffer without probe by pouring over filter. Allow at least a few minutes for

---

1 This is a useful technique and sounds much simpler than it actually is.

2 Hybridization buffers become highly radioactive. Wear gloves for protection.

3 Heat-sealable bags are available where sealing units are purchased. Kapak Corporation's bags are recommended. Their Scotch-pack #405 bags hold 20 × 20 cm NC filters.

4 When filters are wet due to prehybridization (prewetted in dish, not bag), filters are more difficult to maneuver into bag. Alternatively, it is possible to prehybridize inside the bag, cut a corner of the bag to inject more buffer with radioactive probe, and then reseal or clip the bag at the corner.

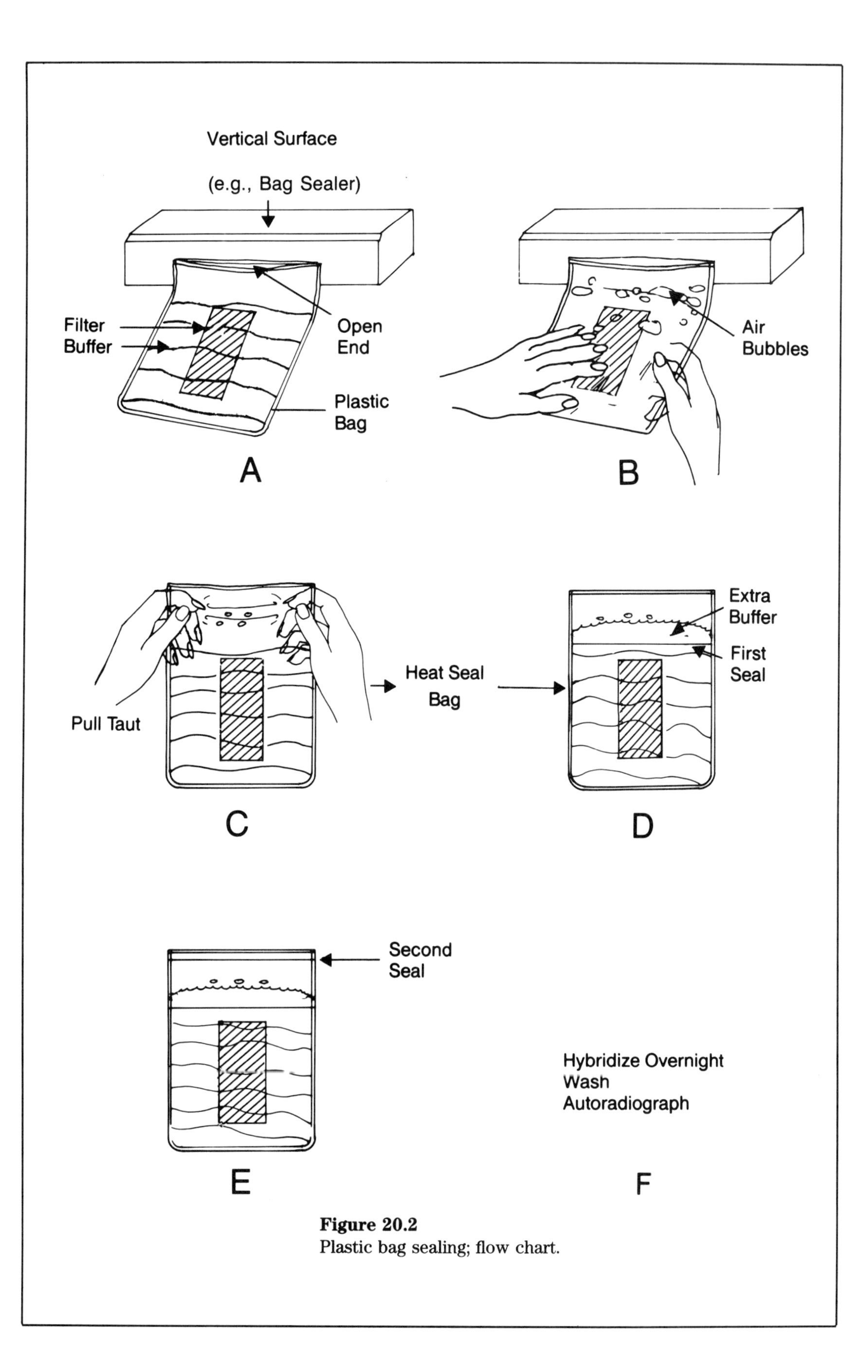

Figure 20.2
Plastic bag sealing; flow chart.

prehybridization and then add additional hybridization mix with probe. Flatten the bag so that all major air bubbles and most small bubbles are removed from the liquid, the filter(s) is uniformly wetted, and the bag is sealed with all liquid inside. Refer to steps in Figure 20.2.

2. Place bag flat on horizontal surface with open end extended up, leaning against a vertical surface.
3. Smooth liquid around filters. Push air bubbles up toward open end.
4. Place hands on either side of bag at top edge of liquid and pull taut while lifting up bag so that air bubbles will "pop" to top edge.
5. Lift bag, still holding side edges, place bag in heat sealer, and make seal near top edge of liquid. There will be a small amount of liquid above the seal.
6. Make second heat seal at top edge of bag to seal in remaining liquid.
7. Bag is now properly sealed and ready for incubation.

SECTION **20-3.**

Optical Density Analytical Measurements

DESCRIPTION

This method is used to estimate the concentration of DNA or RNA by UV absorption spectrophotometry or to estimate the number of bacterial cells in suspension by visible light spectrophotometry. Quartz cuvettes are used for UV spectrophotometry, and plastic or glass ones are used for visible spectrophotometry. An alternative method, the ethidium bromide agarose plate method, is useful when only small quantities of nucleic acid are available; it is presented in Section 11-3, step 5.

TIME REQUIRED

15 min

SPECIAL EQUIPMENT

Spectrophotometer

METHODS

To Measure Concentration of DNA or RNA in Solution

1. Take 5 μl of DNA sample or 4 μl of RNA sample and add to 995 μl of H_2O.[1] Mix. Transfer to spectrophotometer cuvette.
2. Blank spectrophotometer with 1 ml of H_2O.

---

[1] The difference in volumes for DNA or RNA is related to their extinction coefficients. For DNA an O.D. of 1.0 is 50 μg/ml, for RNA or single-stranded DNA an O.D. of 1.0 is 40 μg/ml.

3. Put sample in spectrophotometer[2] and read optical density (O.D.) at a wavelength of 260 nm.
4. Concentration of DNA or RNA, in micrograms per microliter of sample, will be 10 times the O.D. reading. For example, if the O.D. of diluted sample is 0.2 at 260 nm, the original sample contains 2 μg of DNA or RNA per microliter.[3]
5. If desired, to assess the purity of sample, take second reading at 280 nm. Determine ratio of O.D. at 260 nm to O.D. at 280 nm. If ratio is approximately 2, the absorption is probably due to nucleic acids. If ratio is below 1.6, there may be protein or other UV absorbers in the sample, and reextraction with SS-phenol/chloroform and ethanol precipitation (Section 20-1) is advised.

To Measure Bacterial Density during Growth

1. Blank spectrophotometer on growth medium (typically, LB medium).
2. Take 1-ml sample of cells growing in this medium. Measure O.D. at 600-nm wavelength.[4]
3. An O.D. of over 1.0 is beyond log phase.[5] An O.D. between 0.2 and 1 represents cells in log phase growth.[6]

---

[2] Clean cuvettes thoroughly before using.

[3] If DNA or RNA sample contains less than 0.1 μg/μl, O.D. readings will be too low for good reliability. Use a less diluted sample and adjust calculations accordingly.

[4] This is a measurement of light scattering.

[5] Cells may lyse at confluency. Growth can be followed with O.D. measurements to chart growth rate, log phase, and plateau.

[6] An inoculated medium takes a number of hours to reach an O.D. reading of 0.1. Typically, cells double every 30 min in log phase. Many methods call for cells to be in mid-log phase for use (i.e., O.D. is 0.3–0.5).

SECTION

20-4. Photographing Gels or Autoradiograms

DESCRIPTION

This procedure yields a record of DNA or RNA position (bands) on agarose gels prior to processing of gel. The UV light excitation of ethidium bromide (red fluorescence) when bound to DNA or RNA is recorded. X-ray film and polyacrylamide gels can also be photographed by these methods.

TIME REQUIRED

5–10 min

SPECIAL EQUIPMENT

Photographic apparatus (e.g., Polaroid MP-4 with type 57, 3,000 ASA film)
Appropriate filters for ethidium bromide fluorescence (e.g., Wratten #9 or 22A)
UV transilluminating apparatus
X-ray light box

METHODS

To Photograph Gels

1. Place gel on UV light box. Turn on visible focusing light or room light.
2. Add ruler to edge of gel to act as distance marker.
3. Move red filter[1] out of camera path. Set camera to viewing/focus mode.[2] Focus camera on gel or on ruler numbers.

---

1 Red filter is Kodak Wratten 9 or 22A, or equivalent.

2 See instructions with your camera.

4. Put film holder in place behind lens. Return red filter to position in front of lens.
5. Put on goggles to protect eyes from UV light. Turn on UV lamp. Turn off room or focusing lights.
6. Open shutter manually for 2–3 sec.[3]
7. Pull out film. Develop as indicated.

To Photograph X-Ray Film or Polyacrylamide Gel

8. Illuminate film with visible light box and photograph, as in steps 3, 4, 6, and 7, without red filter.

---

[3] Exposure time and aperture setting may vary, depending on film type and speed used as well as the intensity of bands. The correct exposure conditions are determined by trial and error.

SECTION **20-5.**

Autoradiography

DESCRIPTION

Autoradiography is used to detect the location of radioactivity on filters or in gels used in many of the experiments described in this book.

TIME REQUIRED

Varies with exposure; 15 min each to set up or develop autoradiogram

SPECIAL EQUIPMENT

X-Ray film developing system. An automated developer (such as a Kodak X-OMAT) is convenient to save time and trouble and to achieve consistency, but manual development is possible. Details on manual development are available from film suppliers.

X-Ray film cassettes (e.g., Wolf X-Ray Corp.)

Geiger counter

REAGENTS

Autoradiography film, generally X-Ray film (e.g., Kodak XAR-5 or XRP-1)

Image intensification screens (e.g., Du Pont Cronex)

^{14}C or ^{35}S-ink marker

METHODS

General Notes on Autoradiography

1. Typically, the filter or gel containing radioactivity is placed in the dark under unexposed X-Ray film and enclosed in an X-Ray film holder cassette for a

period of time. Following exposure, the film is removed in the dark and developed.

2. For almost all conditions with ^{32}P, an intensifying screen is placed over the X-Ray film, with the film sandwiched between the screen and the radioactive source. With this screen, exposure is at −70°C. The screen will decrease resolution but will enhance the signal strength (up to 10-fold for ^{32}P), making it especially useful with low-level radioactive signals. Where higher resolution is desired, such as with sequencing gels (Sections 16-8 and 16-10), expose film without screen at room temperature.

3. Both XAR-5 and XRP-1 films can be used. XAR-5 is faster but has a larger grain size, whereas XRP-1 has approximately one-third the film speed but has a finer grain.

4. Some highly radioactive samples (e.g., minigels from S_1 probe preparation) require exposure of less than 1 min. Quickly place film over gel (gel is covered with clear plastic wrap)[1] in the dark, with one corner of the film aligned with a corner of the gel. Cover film with glass plate. Wait 20–60 sec. The slower XRP-1 film can be used for this procedure. Quickly remove film and develop.

5. If exposure was made at −70°C, allow cassette to warm to room temperature before opening, to prevent condensation. Alternatively, the cold cassette can be rapidly opened and the film developed quickly, before condensation forms. Do not reload a moist or cold cassette; dry before reusing.

6. Filters and gels can be marked with ^{35}S ink on small adhesive labels. This will appear on exposed film and help with orientation of autoradiogram as well as with identification.[2]

---

[1] Typically, ^{32}P samples are covered with clear plastic wrap (Saran Wrap is recommended). This protects the cassette and intensifying screen from contamination with penetrating radioactivity.

[2] The clear plastic wrap may block a significant amount of the ^{35}S signal. If ^{35}S is used as the radioactive source, do not cover with plastic wrap unless sample is sufficiently radioactive to penetrate plastic wrap.

SECTION **20-6**

Making Plates for Bacterial Growth

DESCRIPTION

This method is used to make hard agar-based culture plates for growth of bacterial colonies. In some cases, for growing bacteriophage plaques, these plates containing bottom agar are overlaid with bacterial cell-containing top agar. Two types of agar-containing media are presented: LB agar for most bacterial cell growth and minimal agar for JM103, JM107, or JM109 stock plates.

TIME REQUIRED

1 hr

SPECIAL EQUIPMENT

Autoclave

REAGENTS

For LB plates

- Bacto-tryptone (e.g., Difco)
- Bacto-yeast extract
- Agar
- NaCl
- **1 M NaOH**
- Ampicillin, 50 mg/ml, sterile filtered
- Tetracyline, 12.5 mg/ml in 50% ethanol, sterile filtered

For LB/ampicillin plates with IPTG and Xgal (for pUC insert selection, described in Section 15-4)

Bacto-tryptone
Bacto-yeast extract
Agar
NaCl
1 M NaOH
Ampicillin, 50 mg/ml, sterile filtered
2% Xgal in DMF
0.1 M IPTG, sterile filtered

For minimal agar

Agar
0.1 M $CaCl_2$ sterile filtered
1 M $MgSO_4$, sterile filtered
Na_2HPO_4
KH_2PO_4
NaCl
NH_4Cl
Thiamine, 100 μg/ml, sterile filtered
20% glucose solution, sterile filtered

METHODS

A. LB agar (with or without antibiotics; with or without IPTG and, Xgal)

a. To make 1 liter for LB agar plates: to 950 ml H_2O add:[1]

10 g bacto-tryptone
5 g bacto-yeast extract
10 g NaCl
Adjust pH to 7.4 with 1 M NaOH
Add 15 g agar
Add H_2O to make 1 liter

Autoclave to sterilize
Cool to 50°C. Add antibiotics (1 ml of a 1,000× stock) if desired.[2]

---

1 The use of pure, deionized H_2O is important. Use only autoclaved glass or sterile plastic dishes.

2 A 1,000× stock of tetracycline (12.5 mg per ml in 50% ethanol) or ampicillin (50 mg per ml of H_2O) is used. Sterilize by filtration and add to medium. The concentrated stocks may be stored in the dark at −20°C.

b. Add 5 ml of 2% Xgal solution and 5 ml of 0.1 M IPTG, if needed (Section 15-4).
c. Pour into dishes on level surface. Use 25-ml aliquots for 90- or 100-mm plates (use glass pipette for transfer). Medium should be approximately 0.3 cm in depth.[3]
d. Let cool to harden.

B. Minimal agar plates
To make 1 liter, weigh

6 g $Na_2HPO_4 \cdot 7H_2O$
3 g KH_2PO_4
0.5 g NaCl
1 g NH_4Cl

Add to 950 ml of H_2O. Dissolve salts and adjust pH to 7.4 with 10 N NaOH. Add 15 g agar. Autoclave the above to sterilize. Cool to 60°C. Add the following sterilized solutions:

10 ml of 20% glucose, filter sterilized
1 ml of 0.1 M $CaCl_2$ autoclaved
1 ml of 1 M $MgSO_4$, autoclaved
50 μl of thiamine, 100 μg/ml

Adjust volume to 1 liter with sterile H_2O. Pour into plates, as in step c above.

C. Storage
Store poured plates, covered to keep sterile at 4°C until ready for use. Do not tape edges or condensation may lead to contamination. Warm to room temperature before using.

---

[3] Cells are usually grown in 90- or 100-mm dishes. Other size dishes can be used; adjust volumes accordingly.

SECTION **20-7.**

Titering and Plating of Phage

DESCRIPTION

When bacterial cells are grown in top agar, they form a cloudy lawn of cells at confluence. If phage are added to this lawn of plating cells, a phage near a cell will infect that cell, eventually lysing the infected cell. Through repeated rounds of infection and lysis, a phage particle can produce 10^7 phage in a very small region, lysing the surrounding bacteria and forming a clear plaque. Each original phage represents a plaque-forming unit (pfu).

It is often desirable to count, or titer, the phage in a given purified sample. To do this, a small aliquot of the sample is plated in a lawn of bacteria. The number of plaques formed is counted, and the pfu per milliliter of the sample is calculated. Often the entire plate does not have to be counted. Representative squares on a plate marked with a grid can be counted, and the number of plaques per volume added to the dish can be calculated from the representative number.

In general, the following steps can be taken to titer and plate phage:

1. Prepare solutions (e.g., LB, maltose, $MgSO_4$, TMG, top agar), as described in Section 13-5. Grow up plating bacteria appropriate for the phage to be titered (e.g., LE392 cells for EMBL3). Inoculate 50 ml of sterile LB medium plus 1 ml of sterile 20% maltose with a single colony of bacteria. Grow cells to an O.D. of 0.5 at 600 nm. Collect cells by centrifugation for 10 min at 3,000 × g. Resuspend pellet in 25 ml of sterile 10 mM $MgSO_4$. These are plating cells and may be stored at 4°C for at least 1 week.
2. Make a series of 10-fold dilutions of 1 μl of bacteriophage in TMG buffer in a series of microfuge tubes. Cover the expected range of bacteriophage titers for your sample. Add 1 μl of diluted phage to a sterile 12 × 75 mm culture tube. Use dilutions predicted to contain a countable number of pfu's in 1 μl volume. The titer of many bacteriophage stock solutions is typically 10^9–10^{10} pfu/ml.
3. Add 200 μl of plating bacteria to each culture tube.
4. Add 2.5 ml of top agar to each tube (cooled to 48–50°C so that bacteria will not be killed).

5. Pour immediately over bottom agar on a 90-mm culture plate. Use plates with grids etched in bottom for easy counting. Make sure that bottom agar plates have warmed to room temperature or 37°C before use.
6. Allow top agar to harden for 15 min at room temperature.
7. Invert plates. Incubate overnight at 37°C.
8. Plaques will be evident. Be sure to count plaques before they are overgrown (i.e., touching each other).
9. Count and calculate number of plaques per dish. Also check number of plaques from control dishes containing plating cells without added phage. Control dishes should not contain plaques.
10. Calculate titer in original samples in pfu per milliliter. This will determine how much of the sample will be required for other procedures as well as the efficiency of library formation (Sections 13 and 14).

SECTION

21

Specialized Methods

SECTION 21-1. Transgenic Mouse Preparation

DESCRIPTION

Transgenic mice are generated by the introduction of a cloned piece of DNA into a fertilized mouse egg, with subsequent growth of the embryo to birth and beyond. Successful incorporation of the DNA into the germ line can later be determined by analysis of genomic DNA in offspring. This method allows the function of specific genes and regulatory elements to be examined in the context of the whole animal with its complex program of development and tissue differentiation.

The first step is to remove freshly fertilized eggs from a female mouse. These embryos are washed and prepared for injection. A special microscope/micromanipulator setup is required. Embryos are individually immobilized by a holding micropipette while cloned DNA pieces are injected into the male pronucleus through a second, injection pipette. Following injection, groups of embryos are placed in the uterine tubes of pseudopregnant female mice. The resulting births are screened for the presence of the cloned DNA within the genome (e.g., by mouse tail DNA preparation and Southern blotting).

At present, this difficult method has been successfully completed in only a handful of laboratories. Be prepared for the many steps of trial and error that will most likely be necessary before the technique is perfected.

To date, a number of functional genes have been introduced into mice by these methods, including oncogenes and the growth hormone gene. By coinserting regulatory sequences along with the coding region of interest, expression of the gene may be regulated. Other transgenic insertions have served to disrupt the host's genes and lead to functional alterations. It is likely that each embryo injected will have the same piece of DNA inserted at random locations in the genome, if at all. Thus, variation in expression can occur. Note: This method requires animal care facilities and knowledge of animal surgery, sterile culture techniques, and micromanipulation of cells.

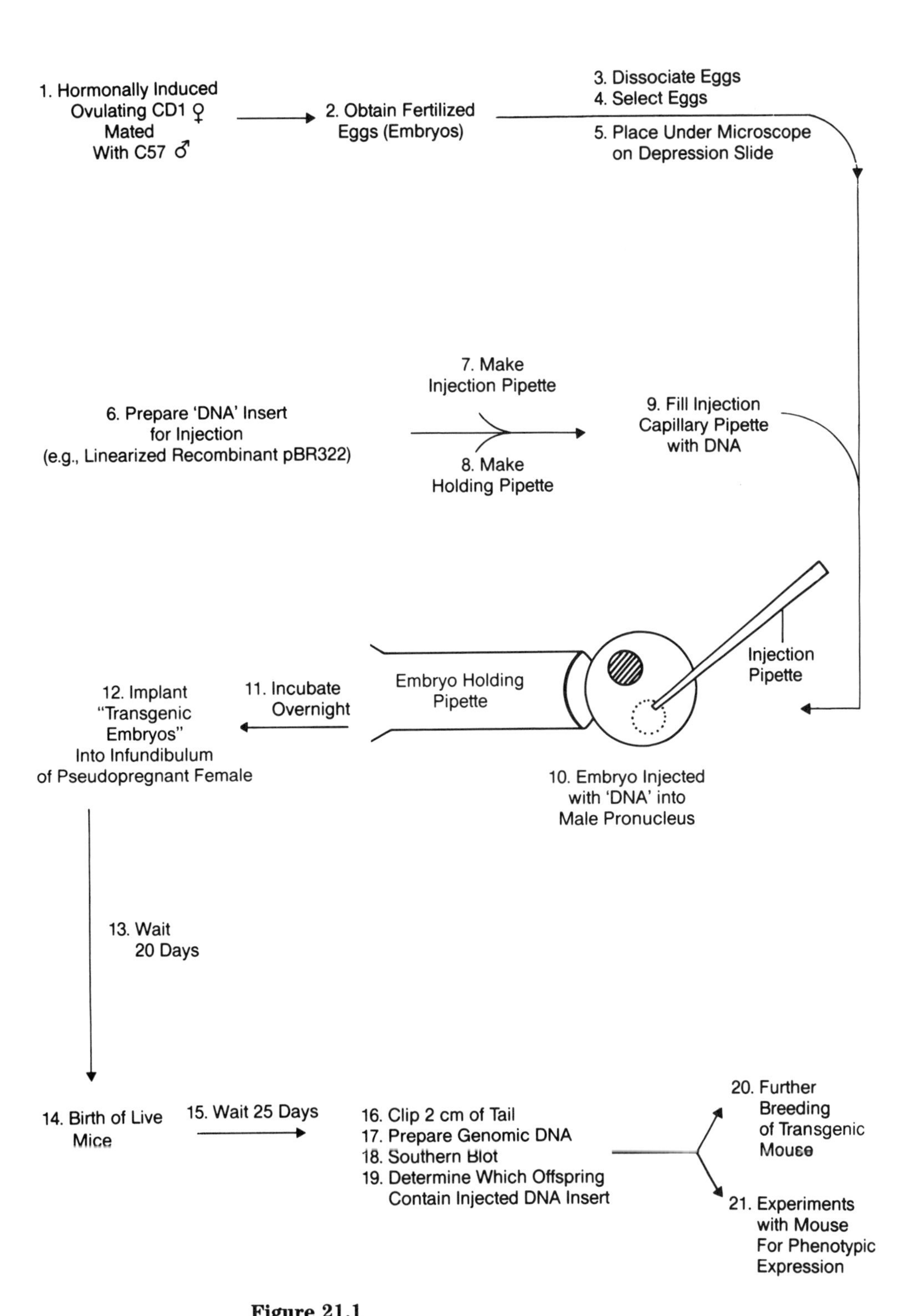

Figure 21.1
Transgenic mouse analysis; flow chart of steps.

TIME REQUIRED

Day 1—4 hr to prepare animals

Day 2—4 hr to prepare embryos
6 hr to inject embryos
4 hr to implant embryos

SPECIAL EQUIPMENT

Inverted microscope, 40× lens with 10× eyepiece and a long working condenser (e.g., Nikon)

Micromanipulators (two) (Leitz), with two single instrument holders and joysticks

Dissecting microscopes, one regular, one illuminating from below with fiberoptic light sources

Pipette puller (vertical) (D. Kopf Inst., Model 700/C)

Compressed air source

Microforge (de Fonbrune)

Surgical instruments including watchmaker's forceps

Controlled environmental chamber

REAGENTS

Whitten's medium (100ml)

NaCl	0.514 g
KCl	0.0356 g
KH_2PO_4	0.0162 g
$MgSO_4 \cdot 7H_2O$	0.0294 g
$NaHCO_3$	0.190 g
Glucose	0.100 g
Ca lactate · 5 H_2O	0.053 g
Na lactate (60% syrup)	0.37 ml
Na pyruvate	0.0025 g
BSA (best quality globulin free)	0.4 g
Na_2EDTA	0.0028 g
HEPES	0.238 g
Penicillin (10000 U/ml)	0.1 ml
Streptomycin (50 mg/ml)	0.1 ml

After each component has been dissolved in sterile, deionized water, sterile filter the medium through a Millipore Sterifil System that was previously rinsed with 10 ml of medium. Pour the sterilized 90 ml of Whitten's medium into a sterile bottle and bubble 5% O_2, 5% CO_2, and 90% N_2 through the solution for 30 min. Store (tightly capped) in refrigerator until use (no longer than 1 week).

Balanced salt solution sterile (e.g., Hank's)

DNA injection buffer, 1 mM Tris-Cl, 0.1 mM EDTA, pH 7.2

Animals

1. Egg donors: CD-1 (e.g. Charles River CD-1 Ha/Icr) female mice, 3–4 weeks old. Allow the mice 1 week to adapt to the light/dark cycle before use.
2. Foster mothers: CD-1 females 8 weeks to 6 months old.
3. Breeding males: C57 Bl/6J males 8 weeks to 1 year old.
4. Sterile (vasectomized) CD-1 males vasectomized at 4 weeks. Wait 1 month, check for sterility, and use until 8–12 months old.

Maintain mice in a temperature/humidity controlled environment with a 14-hr light/10-hr dark cycle. Animals are given food and water ad libitum. With this system, all offspring should not be white, or "sterile" males were not vasectomized.

Drugs

1. Pregnant mare's serum (e.g., Gestyl from Organon or gonadotropin from Sigma). Dilute to 50 U/ml in sterile balanced salt solution. Store aliquots at −20°C.
2. Human chorionic gonadotropin (Sigma). Dilute to 50 U/ml in balanced salt solution. Store aliquots at −20°C.
3. Na Nembutol (stock solution of 2 mg/ml H_2O) (Abbott Labs).

Hyaluronidase; dilute to 50 μg/μl in Whitten's medium for use (Sigma).

Silicone oil (Dow 200, 50 CS viscosity, or methyl silicon, 50 DC 200, William F. Nye, Inc.). For use, autoclave oil and then equilibrate 1 volume of oil against 7 volumes of Whitten's medium. Add 1 μl antibiotics (penicillin/streptomycin) per 1 ml of equilibrated oil. Store in sealed sterile container.

Gas: 5% O_2, 5% CO_2, 90% N_2. Pass through water (washing bottle) for use in environmental chamber.

Antibiotics (penicillin, 10,000 U/ml; streptomycin, 50 mg/ml) (GIBCO).

METHODS

In Advance

Make media. Prepare DNA for injection. Purify DNA after RE digestion to linearize plasmid by agarose gel electrophoresis, and electroelution. SS-phenol/chloroform extract and ethanol precipitate. Dry completely and resuspend at 1 μg/ml in DNA injection buffer. Centrifuge prior to use to remove contaminating particles.[1]

---

1 Use sterile plasticware and other items for this procedure.

Prepare Female Mice for Egg Collection

(Or can use more mice, not timed, and determine positives by checking for vaginal plugs.)

1. Time-ovulate female mice by injection of 5 U pregnant mare's serum (in 0.1 ml, intraperitoneally) approximately 12 hr before the midpoint of the dark cycle.
2. At 48 hr after the first injection, inject 5 U of human chorionic gonadotropin in 0.1 ml.
3. Following injection, house each female overnight with one breeding male.

Next Day

4. Select those female mice with a vaginal plug within 3 hr of onset of light cycle.
5. Sacrifice plugged mice (approximately 16 hr after the midpoint of the previous dark cycle) and remove oviducts to Whitten's medium in sterile culture dish. Medium and dish should be pre-equilibrated to room temperature for a few hours. Transfer to a second dish with Whitten's medium to wash.
6. Tease open the ampulla (transparent bulge in oviduct) with watchmaker's forceps and pull out follicle cells (with fertilized eggs or embryos) surrounded by cumulus covering.
7. Place cumulus/embryos on a drop of Whitten's medium containing 50 μg/ml of hyaluronidase in the middle of a 5-cm culture dish.
8. The hyaluronidase will cause the cumulus to degenerate, and embryos will be released (5–10 min). Monitor cumulus degeneration under dissecting microscope. After this is complete, wait 5 min for egg stickiness to decrease.
9. Transfer embryos by using a modified Pasteur pipette (a glass pipette is pulled over a low flame by hand and cut to a final length of 4 cm with a tip I.D. of 0.5 mm). Use a rubber bulb to control movement in pipette or a clean mouthpiece with tubing for increased control.
10. Transfer embryos in a minimal volume through a series of four culture dishes containing Whitten's medium. Disperse embryos in each transfer to dilute hyaluronidase and other contaminants.[2]
11. Leave embryos in the last culture dish and incubate in the warming chamber described in step 12 until ready for use.
12. The warming chamber (can be an anaerobic jar) is set at 37°C. The chamber is covered and the gas mixture (with 5% CO_2, and passed through water for humidity) continuously flows into the chamber. Keep in this controlled environment for 5–10 min during instrument setup.

---

[2] Washing is done at room temperature. Keep stock of Whitten's medium in a sterile plastic 10-ml syringe at 37°C for use in later steps.

Injection Apparatus

13. a. An inverted microscope fitted with differential interference contrast optics is prepared for use. Micromanipulators are placed on each side of the microscope (adjusted to appropriate height with custom designed aluminum blocks). All components are clamped to a microscope stand resting on a slate block or other heavy table (with proper vibration dampening).
    b. The left-hand micromanipulator is for the holding pipette. The instrument collar is clamped to thick-walled vacuum tubing (3.5 mm I.D.) connected to a 5-ml Hamilton syringe and filled completely with paraffin oil (no air bubbles).
    c. The right-hand micromanipulator controls the injection pipette. The instrument collar is connected with vacuum tubing to a regulatable compressed air source.
    d. The holding pipette is made from capillaries of aluminoborosilicate with 1.5-mm O.D., 1.12-mm I.D. (e.g., Frederick Haer and Co., catalog #30-32-0 with omega dot). Pipettes are pulled on a vertical pipetter puller (e.g., D. Kopf), with tips broken to approximately a 100-μm diameter and fire polished to a blunt end (using a de Fonbrune microforge).[3]
    e. The injection pipette is pulled from the same capillaries on the same apparatus. These are pulled to an O.D. of approximately 1 μm.[4]

14. Fill a holding pipette halfway (from front end) with Whitten's medium by using a slight vacuum through tubing attached to a clean mouthpiece. Seal holding pipette in left-hand manipulator with rubber gaskets. Position holding pipette tip over microscope objective, focus on tip, and lift pipette vertically with manipulator.

15. Introduce 0.5 μl of DNA (1 μg/ml) solution into blunt rear end of the injection pipette using a fine, hand-drawn glass fiber.[5] DNA will be drawn into tip of injection pipette by capillary action. Blow DNA from fiber into pipette tip. Hold vertically (tip up) for 2–3 min to allow DNA to travel to tip. Insert into right-hand manipulator collar, sealing in place with rubber gasket.

16. Transfer embryos to a dish with Whitten's medium. Put a drop of Whitten's medium in well of a depression slide and cover with a drop of oil. Transfer eggs to one side of the drop.

17. Position the depression slide with the embryos on the microscope stage. Lower both pipette tips into drop containing the embryos.[6] In-

---

[3] For convenience, pull about 25 capillaries of each type at each time.

[4] See note 3.

[5] Use same capillary tubes, drawn over open flame.

[6] Adjust holding pipette syringe so with slight back-pressure that no medium flows into or out of pipette. It may take some practice to become proficient at handling manipulator controls.

crease air pressure in the line with the injection pipette to approximately 20 psi.[7]

Injection of Eggs

18. Individual embryos (only those that are fertilized with visible pronuclei and polar bodies) are picked up by tip of holding pipette using gentle, slight back pressure on the attached syringe.[8] Hold so that the male pronucleus is in focus and in the center of the embryo.[9]

19. Position tip of injection pipette directly in line with and above pronucleus.

20. Move injection tip slowly forward into embryo and into pronucleus. A swelling in the pronucleus will be observed since DNA is constantly being ejected from the tip.

21. As soon as swelling is observed, immediately but carefully withdraw injection pipette.[10]

22. Inject entire group of embryos. As each is injected, transfer to the other side of the drop.

23. Remove injected embryos from depression slide with washing pipette.

24. Place embryos in a fresh drop of Whitten's medium to wash, with a slight agitating action.

25. Remove embryos and place in a fresh drop of Whitten's medium in a 10-cm plastic culture dish. Replace dish in controlled environment chamber.

26. Eggs are ready for transfer to foster mothers.

Implanting Eggs in Foster Mothers

27. The day prior to embryo injection, mate foster mothers with sterile (vasectomized) males. Place 9 males overnight with 27 females. Check females for vaginal plugs (there should be about nine plugged animals).

28. Anesthetize pseudopregnant mothers with Nembutol.[11]

29. Dissect to expose ovaries through animal's back. Use dissecting microscope.

30. Gently externalize ovary and oviduct. Make a small tear in ovarian sac with fine forceps.

---

[7] Throughout entire procedure, continual air pressure will eject a stream of DNA out of tip of injection pipette.

[8] Hold embryos firmly, but do not draw into barrel.

[9] See note 8.

[10] Because injection pipette diameter cannot be finely controlled, air pressure may have to be adjusted to give good DNA outflow.

[11] This dose of Nembutol was empirically determined. A typical dose is 35 μl of a 2 mg/ml solution per gram of body weight. Conditions may vary.

31. Transfer pipettes have been previously made by pulling the same capilaries to an I.D. of 150–200 μm and an O.D. of 300–400 μm, 8–10 cm in length. Fire-polish one end in a flame and make a 45°C bend about 1 cm from the polished tip. Connect pipette to rubber tubing and a clean mouthpiece for most accurate transfer.

32. Transfer embryos (5–15 per side) into the infundibulum in a minimal amount of Whitten's solution by carefully controlling air flow to the transfer pipette.

33. Gently replace ovary in body cavity. Suture body wall and skin.

34. Repeat for other uterine tubule.

35. Return foster mother to animal facility for recovery and birth of pups, about 20 days later.

Analysis

36. Analysis of genomic DNA from pup tail pieces will indicate presence of inserted DNA ("transgene") in genome (Sections 5-3, 5-5, and 5-6). If regulatory sequences were also inserted, regulation of a specific phenotype or expressed mRNA is also possible.

REFERENCES*

Brinster, R. L., Chen, H. Y., Trumbauer, M., Senear, W., Warren, R., and Palmiter, R. D., Cell *27*:223, 1981.

Gordon, J. W., and Ruddle, F. H., Meth. Enzymol. *101*:411, 1983.

Hoppe, P. C., and Pitts, S., Biol. of Reproduction *8*:420, 1973.

Mullen, R. J., and Whitten, W. K., J. Exp. Zool. *178*:165, 1971.

Stewart, T., Pattengale, P., and Leder, P., Cell *38*:627, 1984.

Whittingham, D. G., J. Reprod. Fertil. *Suppl. 14*:7, 1971.

---

* Details of this method were provided by Dr. E. Sinn, based on the procedure of T. Stewart.

SECTION **21-2.**

Monoclonal Antibody Production: Hybridoma Fusion

DESCRIPTION

This is an effective technique for fusing antibody-producing mouse spleen cells and immortalized mouse myeloma cells for isolation of the antibody-producing clones with amplification of the antibody product. Following dilution to single-cell aliquots, clones are grown and screened for antibody produced to a specific product. This method allows for the production of highly specific (monoclonal) antibody in quantity for further use.

TIME REQUIRED

4 months from first injection to identification of positive clones

SPECIAL EQUIPMENT

Coulter Counter or other cell counter
Cell culture equipment (including liquid N_2 cell freezer, CO_2 incubator, and biohood)
12-channel pipettor

REAGENTS

(all solutions must be 0.22 μm sterile filtered)

Phosphate buffered saline (PBS)
NH_4Cl buffer: To make 1 liter:
- 2.06 g Tris base (0.017 M)
- 5.87 g NH_4Cl (0.1 M)
- H_2O to 1 liter, pH to 7.2

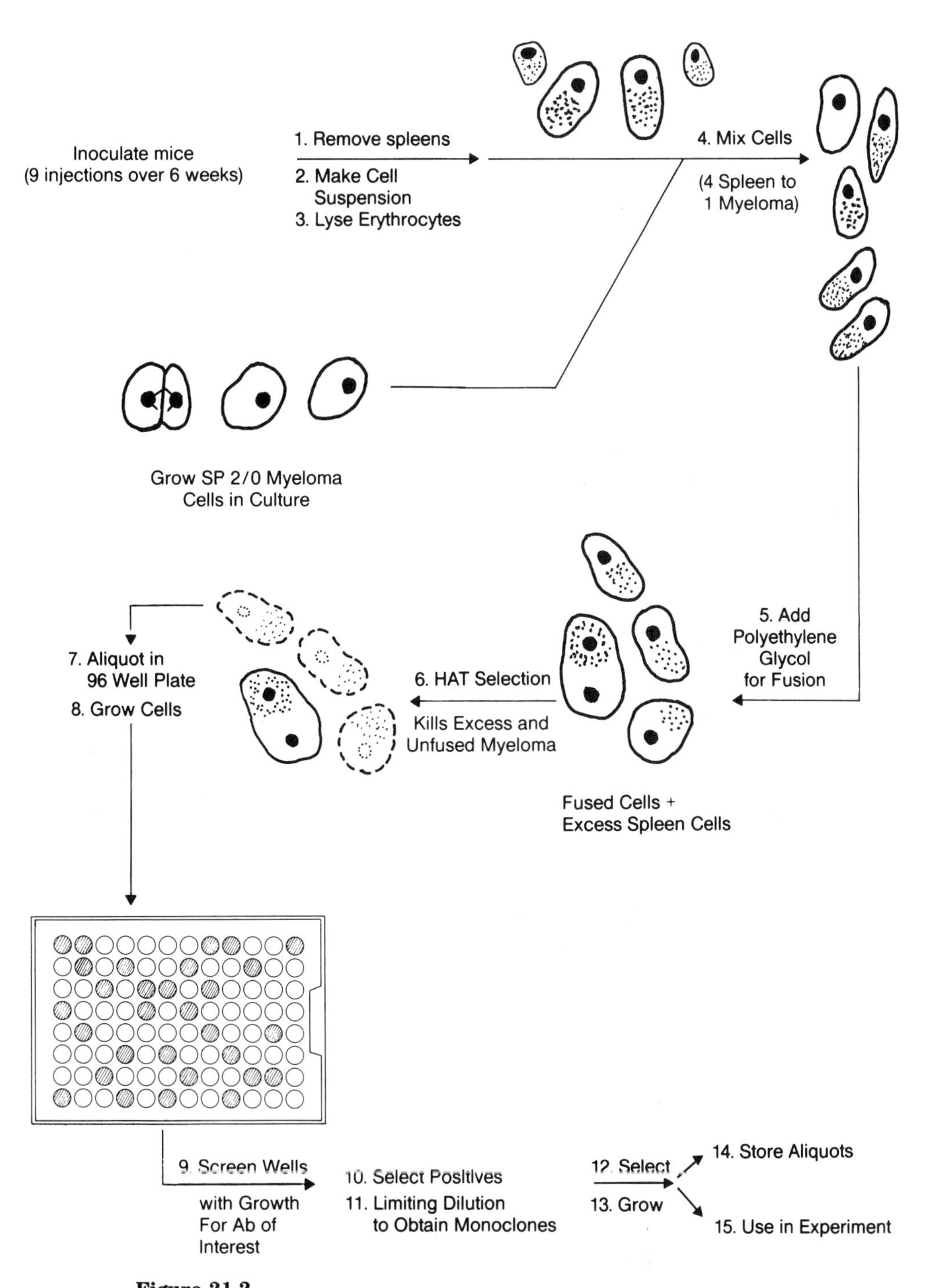

Figure 21.2
Schematic diagram of monoclonal antibody production. Mice are inoculated with antigen of interest. Spleen cells are removed and fused with rapidly-dividing myeloma cells. After dilution, successfully fused cells are screened for the production of antibody of interest.

Dulbecco's modified eagle's medium (DMEM) (Gibco; high glucose)

HT-medium

DMEM (high glucose)	500 ml
200 mM glutamine; use aliquots stored at −20°C	6 ml
HT mix (see below), 100×	6 ml
OPI mix (see below), 100×	6 ml
Fetal calf serum	50 ml
Gentamycin stock (50 mg/ml)	0.6 ml

Polyethylene Glycol 1,000 (50%) (PEG; see note 4)

HT mix (100×)

0.6805 g hypoxanthine in 200 ml H_2O, plus about four pellets of NaOH to dissolve (pH about 10)

0.1937 g thymidine in 200 ml H_2O

Mix both solutions. Add H_2O to make 500-ml sterile filter and store in aliquots at −20°C.

100× aminopterin

1.8 mg aminopterin in 50 ml H_2O

Add 1 N NaOH to dissolve

Bring volume to 100 ml

Sterile filter

Store at −20°C in dark

HAT-selective medium

1 ml of 100× aminopterin

100 ml of HT medium

OPI mix (100×)

7.5 g oxalacetic acid dissolved in 450 ml H_2O

2.5 g pyruvic acid added

Insulin, 1,000 U, dissolved in 0.1 N HCl with vortexing; add to above and H_2O to bring volume to 500 ml.

Make 10- to 50-ml aliquots. Store at −20°C.

Complete and incomplete Freund's adjuvant (Gibco)

BSA (grade V, Miles)

0.15 M NaCl containing 0.05% Tween 20

PBS containing 0.05% Tween 20 and 0.1 mg/ml BSA

Goat anti-mouse IgG conjugated to alkaline phosphatase (Cappel Co. or TAGO, Inc.)

p-Nitrophenyl phosphate (Sigma), 0.1 mg/ml in 10% diethanolamine

METHODS

In Advance

Inoculate mouse with desired antigen. Inoculation will depend on antigen used, availability, purity, and so on. The inoculation can be with nonpure samples, because the monoclonal antibody isolated will be directed against a single epitope. It is also possible to use whole cells or membranes. Pure antigen is needed for the screening assay.

It is best to inoculate at least 2 (and up to 10) mice, injected intraperitoneally with 25–1,000 μg of the antigen in a 1:1 emulsion with complete Freund's adjuvant. Generally, three injections every 2–3 weeks as follow-up boosts with antigen mixed 1:1 with incomplete Freund's adjuvant will work well. The titer or dilution of antisera from each mouse that can be used to identify the antigen should be checked using an ELISA (see below) and should be in a dilution of at least 1:2,000. The serum containing the polyclonal antibodies can be removed from the tail vein or by eyebleed to develop the screening assay and test the titer as it develops in the animal. The intended screen for positive clones needs to be established at this point, with the polyclonal serum, to determine its later utility. The mouse to be used should receive one last intravenous boost with the antigen 3 days before the removal of the spleen. For variations on inoculation or screening techniques, consult the suggested references.

The ELISA Assay

Various screening assays have been developed, but we feel that the enzyme-linked immunosorbent assay (ELISA) works well for many uses. In a typical ELISA, 200 μl of the purified protein (antigen of interest) at 1 μg/ml of PBS is placed in each well of a 96-well plastic plate (e.g., Costar) and incubated for 48 hr at 4°C with gentle rocking. Each well is washed with 0.15 M NaCl containing 0.05% Tween 20. The mouse serum to be evaluated is serially diluted (1:5) in PBS containing 0.05% Tween 20 and BSA at 0.1 mg/ml, to a dilution of at least 1:5,000. Aliquots (200 μl) of each dilution are added in duplicate to wells on the pretreated 96-well plate to determine the optimal titer. In addition, normal mouse serum or preimmune serum will be added as a control. Plates are incubated for 1 hr at room temperature, and wells are washed as noted above. Aliquots (200 μl) of goat anti-mouse IgG conjugated to alkaline phosphatase, typically diluted to 1:400 (dilution must be empirically determined with each lot) in PBS containing 0.5% Tween 20, are added to each well. Plates are again incubated for 1 hr at room temperature. Following incubation, wells are rewashed as noted above and 200 μl of the chromogenic substrate, *p*-nitrophenyl phosphate, is added. Color development usually takes 1 hr, with positive wells turning yellow. The positive responses can be monitored visually or quantitatively with a vertical beam spectrophotometer (or ELISA reader).

Obtain and grow myeloma cells to log phase.[1]

---

[1] Mouse myeloma cells are malignant transformed B lymphocytes. Because they are similar to B cells, this leads to good fusion rates. A number of drug-sensitive myeloma lines are commercially available. The cell line used should be compatible with the injected animals. We use primarily SP2/0 (American Type Culture Collection), be-

1. Place myeloma cells in a sterile 50-ml capped tube.
2. Aseptically remove spleens of mice with a polyclonal titer of over 1:2,000 (from ELISA assay). Place spleens in culture dish containing 5 ml of ice cold DMEM without serum on ice. Force spleens through 150 wire mesh or nylon mesh with forceps until a single cell suspension is formed (about 10^8 cells). Place spleen cells in a second capped 50-ml tube and place on ice for 5 min while debris falls to the bottom. Transfer supernatant to a new tube.
3. Centrifuge both spleen and myeloma cell tubes at 400 $\times$ g for 5 min. Aspirate DMEM off the pellets and wash cells again with 5 ml DMEM.
4. Lyse red blood cells in the spleen cell tube by adding 10 ml of NH_4Cl buffer. Incubate on ice for 8 min. Slowly fill tube with DMEM.
5. While lysing the red blood cells, resuspend myeloma cells to 50 ml with DMEM.
6. Centrifuge both the myeloma and spleen cell tubes at 400 $\times$ g for 5 min. Aspirate off medium from both tubes.
7. Resuspend pellets in each tube to 25 ml with DMEM.
8. Count and check viability of myeloma and spleen cells.[2]

Fusing Cells

9. To the full volume of spleen cells add enough myeloma cells to make a (spleen:myeloma) ratio of between 4:1 and 10:1.[3] Mix gently.
10. Mix cells and centrifuge at 600 $\times$ g for 10 min to form a tight pellet. Aspirate off all supernatant. (Note: Perform all remaining steps, except centrifugations, at 37°C.)
11. Loosen pellet with a slight tap and slowly add 1 ml of 50% PEG[4] over 1 min while stirring gently with the pipette to break up pellet.

---

cause it does not secrete light chain of its own and is a very stable clone. NS1 is also used; it makes, but does not secrete, kappa light chain. Both of these cell lines are sensitive to aminopterin, so nonfused myeloma cells can be killed by growth in HAT, an aminopterin-containing medium. Myeloma cells need to be screened routinely for HAT sensitivity because it can be lost. Cells are grown in HT medium and must be in log growth phase before use. They are split 1:2 each day for 3 days prior to fusion.

[2] Counts can be made with a hemocytometer or Coulter counter. Viability can be determined by trypan blue exclusion: Mix 1 drop of trypan blue with aliquot of cells. Examine under a microscope. Viable cells exclude stain; nonviable cells are blue.

[3] More spleen cells than myeloma cells are needed.

[4] The time, type, and amount of PEG must be determined empirically. We have found that 50% PEG-1,000 works best; others may use different percentages and types. To determine which works best, "sham fusions" can be performed by fusing spleen cells with myeloma cells and checking the fusion rate (30–50% is good). Cells are fragile after PEG treatment, and timing is also of critical importance for this step.

12. Stir with pipette tip for another minute.

13. Add 1 ml of DMEM with stirring over 1 min.

14. Add an additional 1 ml of DMEM with stirring over 1 min.

15. With 10-ml pipette, slowly stir in 7 ml of DMEM over 2 min.

16. Centrifuge at 600 × *g* for 10 min at room temperature. Aspirate off medium.

Selection and Screening

17. Add 10 ml of HAT-selective growth medium by aiming directly at pellet. Do not pipette up and down to mix.

18. Add additional HAT-selective medium to make a total of 10^6 cells per milliliter. Allow to sit briefly until cells disperse.

19. Aliquot 100 μl of cells per well in 96-well tissue culture plates (approximately 100,000 myeloma cells per drop).

20. Allow clumps of cells to settle in the tube while aliquoting drops. After suspended cells are aliquoted, attempt to break up clumps by up-and-down action of pipette and then pipette the remainder of cells and clumps into additional wells.

21. Add an additional 55 μl of HAT-selective medium per well (with a 12-channel pipettor).

22. Incubate plates at 37°C with 5% CO_2 for 4 days.[5]

23. Carefully remove upper two-thirds of medium (about 100 μl) and feed clones with 200 μl of HT medium (no aminopterin)[6] per well.

24. Feed as needed, approximately weekly, until clones appear.

25. Test those aliquots that formed yellow supernatants (due to pH change with growth) for specific antibody production. The screen depends on the properties and availability of the starting antigen.[7]

---

[5] The aminopterin in HAT medium will kill the nonfused myeloma cells within 24 hr. We allow 4 days to be safe. Nonfused spleen cells will not divide. Therefore, cells growing in clones in the wells are fusion products. Supernatants in these wells will turn medium yellow, indicating growth of surviving cells. Yellow wells are tested. Although these clones may be monoclonal, a series of limited dilutions (step 29) will ensure this result.

[6] Feed by carefully aspirating top two-thirds of liquid in well and replacing with fresh HT medium (approximately 200 μl).

[7] Screening is based on the antigen, its properties, and what is required for its use. This is the most difficult part of antibody production. The screen will be radioimmunoassay (RIA), ELISA, or a functional test. This depends on the antigen and the testing systems available, as well as the desired function(s) of the antibody produced. The screen should be developed before fusion is initiated, typically during the immunization period.

26. Expand each positive clone to 5 wells of a new 96-well plate or 1 well in a 24-well plate. Feed and grow until yellow. Test again for continued positivity.
27. Expand each positive clone in 25 cm^2 flasks with 5 ml of HT medium.
28. Freeze several vials of each as soon as possible. One milliliter of cells is frozen in DMEM, 10% serum, and 10% dimethylsulfoxide (DMSO) in a cell freezer (or slowly to liquid N_2) at 10^7 cells per milliliter and stored at $-70°C$.
29. Clone by limiting dilution. Dilute so that there will be an average of one cell per 2 wells on a 96-well plate. There should thus be 30–50 positive wells per plate. If more positive wells arise, begin limiting dilutions again (more dilute). Statistically, two to three successful limiting dilutions are required of the original clone to ensure its monoclonality.
30. Grow and freeze positive clones. Monoclonal antibodies can now be used as probes for the antigen of interest.

REFERENCES*

Davis, L. G., Manning, R. W., Callahan, A. M., Wolfson, B. Baldino, F., Jr., and Toy, S. T., J. Neurosci. Meth. *14:*15, 1985.

Engvall, E., Meth. Enzymol. *70:*419, 1980.

Goding, J. W. (ed.), *Monoclonal Antibodies: Principles and Practices.* Academic Press, New York, 1983.

Kennett, R. H., McKearn, T. J., and Bechtol, K. B. (eds.), *Monoclonal Antibodies.* Plenum Press, New York, 1983.

Kohler, G., and Milstein, C., Nature *256:*495, 1975.

Oi, V. T., and Herzenberg, C. A. In: B. B. Mishell and S. M. Shigii (eds): *Selective Methods in Cellular Immunology.* W.H. Freeman, San Francisco, 1980.

Voller, A., Bidwell, D. E., and Bartlett, A., *The Enzyme Linked Immunosorbent Assay (ELISA).* Borough House, Guernsey, England, 1979.

---

* M. M. Kelley and M. C. Cohen, in personal communications, supplied many of the methods described in this section.

SECTION **21-3.**

In Situ Hybridization of Labeled Probes to Tissue Sections

DESCRIPTION

This method allows the identification of mRNA within intact tissues at both regional and cellular levels. Synthetic oligonucleotide probes, about 25–50 bases in length, are ideal because they can penetrate tissue easily and are sufficiently specific to recognize a single mRNA species. Probes of this length can be designed to be complementary and specific to unique segments of mRNA, and may thus distinguish between mRNAs differing by only a single base. This method has proven useful in the study of tissues containing heterogeneous cell types, such as brain, but is also applicable to many other tissues. The method presented here is for rat brain tissue, and minor modifications may be necessary for other tissues. This technique permits the study of mRNA regulation in specific tissues or cells. Note: This method requires animal facilities and knowledge of animal perfusion and dissection.

TIME REQUIRED

Day 1—2 hr to perfuse animal

Day 2—6 hr to section, prepare, and hybridize tissue

Day 3—4 hr to wash tissue and prepare for autoradiography

SPECIAL EQUIPMENT

Cryostat, freezing microtome or vibratome

Perfusion apparatus

REAGENTS

Phosphate-buffered saline (PBS)

4% paraformaldehyde in PBS

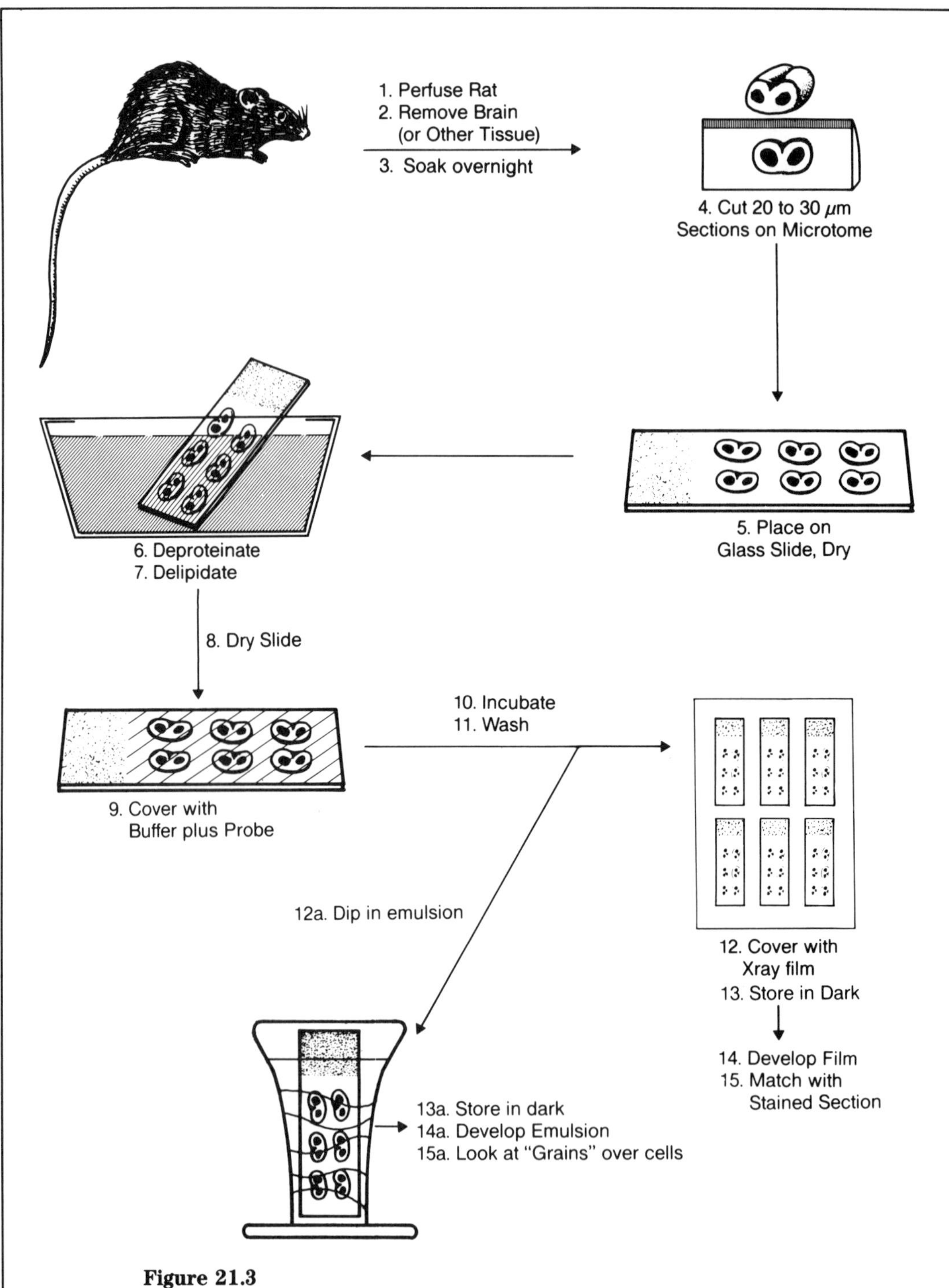

Figure 21.3
In situ hybridization. Tissue of choice (rat brain, in this example) is sectioned and mounted on microscope slides. The sections are protected from RNA degradation, deproteinated, delipidated and hybridized with a radiolabelled nucleotide probe to determine presence of RNA of interest. Hybridized samples can be dipped in emulsion for cellular localization, or autoradiographed on film for regional localization.

Sucrose
Diethylpyrocarbonate (DEP), 0.02% (50 μl in 250 ml autoclaved H_2O)
Proteinase K mixture (makes 100 ml)
    20 mM Tris, pH 7.4
    2 mM $CaCl_2$
    Proteinase K, 1 mg/ml
Prehybridization buffer
    50% deionized formamide (concentration may vary)
    0.1% salmon sperm DNA
    2× SSC buffer
    10× Denhardt's solution
    0.1% SDS

Ethanol (100, 95, 70, and 50%)
2× SSC buffer

METHODS

In Advance

Radiolabel synthetic DNA probe by end labeling (Section 6-2). Labeled probes can be generated by other methods (e.g., DNA from M13 or RNA from SP6). Make subbed microscope slides.[1]

Prepare Tissue

1. Perfuse anesthetized adult rats transcardially with cold (4°C) PBS until blood clears (5–20 min), followed by a perfusion with 4% paraformaldehyde in PBS (about 20 min).[2]
2. Remove brains. Post-fix for 1 hr in 4% parafomaldehyde in PBS at 4°C. Incubate overnight at 4°C in PBS containing 15% sucrose.
3. Dip subbed microscope slides briefly in 0.02% DEP and allow to air-dry.

---

[1] Subbed slides are made by the following treatment:

a. Rinse slide in hot, soapy H_2O. Rinse well with H_2O.
b. Soak in 8% nitric acid for 30 min. Rinse well with H_2O. Air-dry slides.
c. Heat 750 ml of H_2O to 55°C. Add 1.87 g of gelatin and dissolve. Cool to room temperature. Add 187 mg of chromium potassium sulfate.
d. Dip slides three times slowly in the above solution.
e. Dry slides vertically (dust free). Subbing solution will last for up to 1 week at 4°C and subbed slides will last indefinitely in an airtight box.

[2] Perfusion methods are described in many textbooks that cover the handling of laboratory animals. Other fixatives, such as periodate-lysine-paraformaldehyde in PBS, can be substituted. For some tissues, postfixation of unperfused material is effective.

4. Freeze brains in liquid nitrogen. Cut frozen whole brains, or brain area of interest, in 5- to 30-μm sections on a cryostat at -15°C or on a freezing microtome.
5. Place sections on slides.[3] Carefully flatten sections on slide with a fine paintbrush while warming underside of slide.
6. Incubate slides in proteinase K mixture for 15 min at 37°C.[4]
7. Delipidate tissues by dipping sequentially for 5 min each in H_2O twice, once in DEP (from step 3), 50, 70, 95, and 100% ethanol, and then back through this sequence to H_2O again and then in DEP solution.
8. Place slides in prehybridization buffer for 1 hr at room temperature.
9. Make hybridization buffer. A 50-μl volume will be used for each section. Add radioactive probe to make $0.5–2.0 \times 10^5$ cpm per 50 μl of prehybridization buffer. Drain off prehybridization buffer from slides. Add hybridization buffer containing probe in 50 μl "puddles" over each section.
10. Incubate slides overnight in a humidified chamber on a level surface at 4–45°C.[5]
11. Rinse slides in fresh 2× SSC, changed every 15 min for 3 hr at room temperature or higher.[6]
12. Air-dry slides.
13. Visualize labeled areas by autoradiography. Cover with clear plastic wrap, X-ray film (XAR-5), and enhancing screen. Expose for 3–7 days at -70°C.[7]

---

[3] Make sure slides are well marked as to samples. Marking should not be soluble in H_2O or ethanol. A pencil is recommended.

[4] Alternatively, incubate in 0.1 M HCl at room temperature.

[5] Temperature is approximate. The temperature, SSC concentration, and percentage of formamide in the buffer may be varied according to the probe length and stringency required (see Section 6-3). If the probe is shorter than 50 bases, it may be necessary to reduce the hybridization temperature or the percentage of formamide.

[6] Use autoclaved glassware and sterile H_2O, and wear gloves to reduce the chance of RNase contamination. If the background is too high, wash slides again in more dilute SSC buffer and/or at high temperatures or repeat the experiment with less probe. If no signal is detected, increase the radioactive probe concentration, or autoradiograph for a longer time, or hybridize and wash less stringently (e.g., in 4× SSC buffer).

[7] For cellular localization studies, slides can be dipped in photographic emulsion (e.g., a 50% dilution of Kodak NTB2, NTB3, or Ilford L4) and the emulsion developed. The following procedure can be used for dipping slides in emulsion:

a. Make a 1 : 1 dilution of Kodak NTB2 or NTB3 by adding equal volumes of melted emulsion at 41°C and H_2O at 41°C in total darkness (or under a sodium vapor safelight). Avoid the introduction of air bubbles.

b. Dip a few blank slides in emulsion to remove air bubbles and verify the evenness of the emulsion mixture by holding the slide near the safelight.

c. Dip two slides (back to back) into the emulsion. Slowly remove slides and touch the end to the outside of the beaker to remove excess emulsion. (See footnotes d., e. and f. on facing page).

REFERENCES

Angerer, R. C., Cox, K. H., and Angerer, L. M. In: *Genetic Engineering, Volume 7* (J. K. Setlow and A. Hollaender, eds.), Plenum Press, New York, 1985, pages 43–65.

Davis, L. G., Arentzen, R., Reid, J. M., Manning, R. W., Wolfson, B., Lawrence, K., and Baldino, F., Jr., Proc. Nat. Acad. Sci. U.S.A. *83:*1145, 1986.

Gee, C. E., and Roberts, J. L., DNA *2:*157, 1983.

Lewis, M. E., Sherman, T. G., and Watson, S. J., Peptides *6(suppl 2):*75, 1985.

Wolfson, B., Manning, R. W., Davis, L. G., Arentzen, R., and Baldino, F., Jr., Nature *315:*59, 1985.

---

d. Place slides in a vertical rack (e.g., a scintillation vial tray with absorbent paper) within a humidified chamber. Allow emulsion to dry for 2 hr in total darkness.

e. Place slides in a slide box with a desiccant (e.g., Drierite in packs). Tape sides to prevent light leakage. Store at 4°C for 2 days to 8 weeks depending on mRNA abundance and isotope used.

f. Warm slide box to room temperature. Unseal box. Develop emulsion in Kodak D19 developer (1:1 dilution) for 2 min at 16°C, followed by stopping in water (10 sec), fixation (3 min in Kodak rapid fix), and washing in water for 45 min. Dehydrate through ascending concentrations of ethanol followed by xylenes. Coverslip with Permount and examine under microscope.

SECTION **21-4.**

Cloning into Yeast

Although a full description of the methods relating to cloning of DNA into yeast is beyond the scope of this book, some general notes about these methods may prove helpful.

The mammalian cell and bacterial systems presented in this book are used by a majority of molecular biologists and are usually satisfactory for most cloning needs. The use of yeast as a host and as a cloning vehicle has certain advantages, however, and should be noted. Simply put, yeast is a rapidly dividing cell that is easily manipulated and has many of the molecular attributes of eukaryotic cells.

The bacterial hosts described herein have the advantage of containing a simple genome and have good amplification and selection properties. However, they are prokaryotic cells and cannot perform a number of functions seen in eukaryotic cells. For example, bacterial host cells do not glycosylate proteins in the same way as eukaryotic cells can. Also, bacterial cells do not have nuclei, nor do they undergo mitosis and meiosis, making them unsuitable for studying gene expression during the cell cycle.

The mammalian cell methods described in this book allow for the expression of cloned DNA in eukaryotic cells, but these cells have a complex DNA genome and do not have the optimal amplification and selection properties of microbes. Fermentation processes involving yeast have been used by industry for centuries. More recently, the biotechnology industry has turned to yeast host cell systems for the production of protein products not possible in bacterial systems. If one wishes to have the advantages of working with microbial host cells while examining eukaryotic DNA properties, then the use of yeast may be advantageous.

Yeast host cells can also be used to study the expression, regulation, and function of specific genes. For these experiments, one advantage of using a yeast host cell system is that it is possible to make specific changes in the endogenous yeast alleles. In mammalian cell hosts, an exogenous DNA construct is randomly integrated into the genome. In yeast, it is possible to replace a genomic allele with a cloned yeast allele that has been modified in the laboratory because integration of exogenous DNA usually occurs by homologous recombination. Thus, a new gene can be constructed and examined in its normal chromosomal environment. Other advantages of yeast are their easy growth, high degree of characterization, and very high levels of mRNA.

There are also potential limitations to the use of yeast host systems. Yeast does not have infecting viruses, so viral vectors cannot be used with yeast for transformation. Another limitation is that, in general, yeast may not correctly splice out introns from the DNA of higher eukaryotes, thus preventing proper processing of RNA and translation of protein products in some cases.

The most frequently used yeast hosts in molecular biology are strains of *Saccharomyces cerevisiae*, and there are many plasmid systems available for use in yeast cloning. These plasmids often contain at least part of an *E. coli*-based plasmid, including an origin of replication and a selectable marker. As such, they can be transformed into both bacterial and yeast host cells, and have thus been termed *shuttle vectors.* The yeast-integrating plasmids (YIp) are incapable of autonomous replication and are therefore maintained only after insertion into a chromosome by homologous recombination. The yeast-replicating plasmids (YRp), which contain yeast chromosomal DNA that allows autonomous replication of the plasmid, and the derivatives of the natural plasmid from yeast (YEp) give a much higher frequency of transformation than do the YIps. The YIp vectors yield only a few transformants per microgram of DNA, whereas the YRp and YEp vectors yield between 10^2 and 10^5. Typically, the YEp and YRp plasmids are unstable; this property can be used advantageously in some circumstances but may be a disadvantage in others. The YEp plasmids replicate at a high copy number (e.g., about 50 per cell) and can therefore be used to increase the gene copy number of cloned DNA. Lastly, the YRp plasmids can carry a yeast centromere and, in this conformation, be stably maintained in low copy number.

It is also possible to introduce linear fragments of DNA directly into yeast chromosomes by homologous recombination. This is done without introducing exogenous plasmid sequences. Also, by this method, cloned DNA can be introduced to the yeast at predetermined and well-defined locations on the chromosome.

As with other systems, it is possible to add inducible promoters to cloned DNA fragments in order to regulate the expression of the inserted DNA. For example, the *gal* promotor can be added, and is "off" in the absence of galactose and "on" at high levels (i.e., 1,000-fold) in the presence of galactose without glucose. In addition, signal sequences specifying secretion can be added to the inserted DNA fragment. For example, the signal sequence for the mating hormone α-factor allows the efficient export of proteins to the medium.

The genetics of yeast are well characterized. This makes it possible to use yeast as a host for the isolation and selection of genes from heterologous systems by complementation of well characterized mutations in the analogous yeast genes. In addition, it is sometimes convenient to isolate a particular gene of interest from yeast and to use that gene as a heterologous hybridization probe to isolate related genes from other species.

Many of the methods used for bacterial host cells can be applied to transformed yeast cells. For an overview of the methods involved with yeast host systems, see the Strathern et al. reference below.

REFERENCES

Beggs, J. D., Nature *275:*104, 1978.

Lewin, B. (ed.), *Gene Expression*, Vol. 2: *Eukaryotic Chromosomes.* John Wiley, New York, 1980.

Spencer, J. F. T., Spencer, D. M., and Smith, A. R. W. (eds.), *Yeast Genetics.* John Wiley, New York, 1983.

Strathern, J. N., Jones, E. W., and Broach, J. R., *The Molecular Biology of the Yeast Saccharomyces*, 2 vols. Cold Spring Harbor, New York, 1982.

APPENDIX I

Stock Solutions

Common stock solutions, reagents, and chemicals used in the methods contained in this manual are listed below. In some cases, such as with the enzymes and some special chemicals, suppliers are listed after the compound's name. In general, the first name listed is the one used successfully in our laboratories. However, this should not be taken as an endorsement, only as a suggested starting point. See Appendix III for the addresses of certain suppliers. Note that in "Reagents" sections throughout the book, each stock solution is printed in *italics* if it is used straight from the stock bottle, and in **boldface** if it is used as a dilution of a stock solution. Stock solutions are listed in alphabetical order for all buffers. Cell growth media are listed under "Media." The "Reagents" section of each method notes any variations from the stock solution. Some stock solutions are made in concentrated form and are so designated (e.g., 10× or 100×). Note that H_2O in any formula indicates the use of deionized, distilled water. The following stock solutions used in this manual are listed below:

10 M ammonium acetate
10% bromophenol blue
1 M $CaCl_2$
100× Denhardt's solution
DNA, salmon sperm, 10 mg/ml
1 M DTT
0.5 M EDTA
Ethidium bromide, 10 mg/ml
Formamide, deionized
Gel loading buffer
1 M HCl
Hybridization buffer-N (nick)
Hybridization buffer-S (synthetic)
100 mM IPTG
10× kinase buffer
10× ligase buffer
Media for culture
- DMEM
- LB agar
- LB medium

1 M $MgCl_2$
1 M $MgSO_4$
10× polymerase buffer
RSB buffer
3 M sodium acetate
5 M NaCl
10 M NaOH
10% SDS
20× SSC buffer
20× SSPE buffer
SS-phenol
40× TAE buffer

20× TBE buffer
TE buffer
TES buffer
TMG buffer
10% trichloroacetic acid
2 M Tris, pH 7.4
1% Triton X-100
tRNA from yeast, 10 mg/ml
10% Xgal solution
10% xylene cyanol

STOCK SOLUTIONS

Ammonium acetate, 10 M:
Mix 385.4 g of ammonium acetate with 150 ml of H_2O. Bring up to a total volume of 500 ml.

Bromophenol blue, 10%:
Mix 5 g of bromophenol blue in 50 ml of H_2O. Store at room temperature.

$CaCl_2$, 1M:
Dissolve 147 g of $CaCl_2 \cdot 2H_2O$ in 1 liter of H_2O. Autoclave to sterilize. Store at room temperature.

Denhardt's solution, 100×:
10 g polyvinylpyrrolidone (PVP)
10 g bovine serum albumin (BSA)
10 g Ficoll 400
H_2O to 500 ml
Sterile filter. Store at 4°C. Mix before using.

DNA, Salmon sperm (alternatively called *herring sperm DNA*), 10 mg/ml in H_2O; you can also keep 2 and 5 mg/ml stocks. Cut desired amount with a clean scissors and shear by passing solution once forcefully through a 23-gauge needle. Store at 4°C (Sigma, Na salt of DNA from salmon testes, type III).

DTT (dithiothreitol). 1 M stock:
3.09 g in 20 ml of H_2O. Store aliquots at −20°C.

EDTA (ethylenediamine tetraacetic acid, disodium salt), 0.5 M:
To 300 ml of H_2O in beaker add:
93.05 g EDTA-$Na_2 \cdot 2H_2O$
Mix well. Add 10 N NaOH to pH 8.0.
(Will not go into solution until about pH 7.)
Add H_2O to 500 ml.

Ethidium bromide solution, 10 mg/ml. Used to visualize DNA; emits fluorescence when excited at 250–310 nm. M.W. 394.32. Add 0.2 g of ethidium bromide to 20 ml of H_2O. Mix well. Store at 4°C in dark (brown or foil-covered bottle). Handle with gloves and avoid inhalation.

Formamide, deionized. Add 50 ml of formamide to 5 g of a mixed-bed resin (e.g., Bio-Rad AG 501-X8) Stir gently at 4°C and filter twice through Whatman No. 1 filter paper. Aliquot and store at −20°C.

Gel loading buffer. Used to load DNA samples on gels:

To make	10 ml	final conc.
Glycerol	5 ml	50%
40× TAE buffer	250 μl	1×
Bromophenol blue, saturated	1 ml	1%
Xylene cyanol, 10% suspension	1 ml	1%
H_2O	2.75 ml	

Typically, make 10 ml, divide into 1-ml aliquots, and store at −20°C.

HCl, 1 M Stock:

To make 1 liter, add 86.2 ml of concentrated HCl to 913.8 ml of H_2O.

Hybridization Buffer-N. For use with probes from nick translation:

To make approximately 800 ml:

Dextran sulfate	80 g
H_2O	300 ml
Formamide (deionized)	320 ml
20× SSC buffer	160 ml
Tris, 2 M, pH 7.4	8 ml
100× Denhardt's solution	8 ml
Salmon sperm DNA (2 mg/ml)	8 ml

It may be necessary to stir overnight and filter through Whatman No. 1. Store at −20°C. Will last for months.

Hybridization buffer-S. For use with synthetic probes:

Make fresh:

3 ml 10% PVP (polyvinylpyrrolidone)

3 ml 10% BSA

3 ml 10% Ficoll 400

2.2 ml salmon sperm DNA stock (10 mg/ml)

15 ml 20× SSC buffer stock

250 mg yeast RNA

H_2O to 150 ml

IPTG (isopropyl-β-D-thiogalactopyranoside) 100 mM in H_2O:

23.8 mg IPTG in 1 ml H_2O.

Kinase buffer, 10×	To make 5 ml:
0.5 M Tris, pH 7.4	1.25 ml of 2 M stock
0.1 M $MgCl_2$	0.5 ml of 1 M stock
50 mM DTT	0.25 ml of 1 M stock
	3 ml of H_2O

Ligase buffer, 10×, for T4 DNA ligase:

	To make 2 ml:
Tris, 0.5 M, pH 7.4	0.5 ml of 2 M stock
$MgCl_2$, 0.1 M	0.2 ml of 1 M stock
DTT, 0.2 M	0.4 ml of 1 M stock
ATP, 10 mM	0.2 ml of a 100 mM stock, pH to 7.0 with NaOH
BSA, 50 μg/ml	100 μg
	H_2O to make 2 ml

Media for culture: See also individual sections where used.

LB agar (Section 20-6):

To 950 ml of H_2O add:

10 g bacto-tryptone (e.g., Difco)
5 g bacto-yeast extract
10 g NaCl

Adjust to pH 7.4 with 1 N NaOH
Add 15 g agar (8 g for top agar)
Add H_2O to make 1 liter
Autoclave to sterilize
Add antibiotics, if desired, after cooling to 50°C

LB (Luria-Bertani) medium (LB broth; Section 8-1):

To 950 ml H_2O add:

10 g bacto-tryptone
5 g bacto-yeast extract
10 g NaCl

Adjust to pH 7.4 with 1 N NaOH
Add H_2O to 1 liter
Autoclave to sterilize

$MgCl_2$, 1 M:

Add 20.3 g of $MgCl_2 \cdot 6H_2O$ to a final volume of 100 ml of H_2O
Autoclave to sterilize

$MgSO_4$, 1 M:

Add 24.6 g of $MgSO_4 \cdot 7H_2O$ to a final volume of 100 ml of H_2O
Autoclave to sterilize

Polymerase buffer, 10×:

	To make 10 ml:
70 mM Tris, pH 7.4	0.35 ml of 2 M stock
500 mM NaCl	1 ml of 5 M stock
70 mM $MgCl_2$	700 μl of 1 M stock
	7.95 ml of H_2O

RSB buffer (Tris/NaCl/EDTA):

	To make 100 ml:
10 mM Tris, pH 7.4	0.5 ml of 2 M stock
10 mM NaCl	0.2 ml of 5 M stock
25 mM EDTA, pH 7.4	5.0 ml of 0.5 M stock
	94.3 ml of H_2O

Salmon sperm DNA; See DNA, salmon sperm

Sodium acetate, 3 M:

40.8 g of sodium acetate · $3H_2O$ in a final volume of 100 ml of H_20. pH adjusted with acetic acid, as indicated; typically 7.0 for use with DNA, and 6.0 and sterile filtered for RNA.

NaCl, 5 M:

292.2 g of NaCl in a final volume of 1 liter of H_2O

NaOH 10 M. Handle carefully:

200 g of NaOH in 450 ml of H_2O. Mix. Bring volume to 500 ml.

SDS (sodium dodecyl sulfate), 10%.

100-g bottle brought to 1 liter with H_2O.

Use whole bottle to avoid breathing dust.

SSC Buffer 20X:

	To make 1 liter:	4 liter:
NaCl, 3 M	175.3 g	701.2 g
Na_3 Citrate · $2H_2O$, 0.3 M	88.2 g	352.9 g
H_2O	800 ml	3200 ml
Adjust pH to 7.0 with 1 M HCl		
Add H_2O to	1 liter	4 liter

SSPE buffer 20×. Used in plaque lifts:

	To make 1 liter:
NaCl (3 M)	175.3 g
NaH_2PO_4 · H_2O (0.2 M)	27.6 g
EDTA-Na_2 (0.02 M)	Add 40 ml of 0.5 EDTA stock solution
	Add H_2O 800 ml

Add 10 N NaOH to pH to 7.4

Add H_2O to make 1 liter

Sterilize by autoclaving; all components will go into solution in autoclave.

SS-phenol (salt-saturated phenol):

1-lb bottle phenol (e.g., Mallincrodt)
100 ml 2 M Tris, pH 7.4
130 ml H_2O
Heat at 37°C carefully until phenol is dissolved.
Remove upper aqueous phase.
Add: 100 ml 2 M Tris, pH 7.4
25 ml m-cresol (e.g., Kodak)
1 ml β-mercaptoethanol
500 mg 8-hydroxyquinoline

Store aqueous and phenol layers together in hood at room temperature. Use the yellow phenol solution for extraction. Note that the phenol will remain as the top layer in the stock bottle for easy removal. When used to extract DNA, the SS-phenol will be typically the bottom layer.

TAE buffer, 40×:

	To make 1 liter:
Tris base (1.6 M)	193.6 g
Na acetate · $3H_2O$ (0.8 M)	108.9 g
EDTA-Na_2 · $2H_2O$ (40 mM)	15.2 g
pH to 7.2 with acetic acid	
H_2O to make	1 liter

TBE buffer, 20×:

	To make 1 liter:
Tris base (1 M)	121 g
Boric Acid (1 M)	61.7 g
EDTA-Na_2 · $2H_2O$ (20 mM)	7.44 g
H_2O to make	1 liter

Store at room temperature. Remake if major residues appear in solution.

TCA (Trichloroacetic acid) stock, 10%:

Add 227 ml H_2O to a bottle containing 500 g TCA. This is 100% TCA. Dilute to 10% for working stock.

TE buffer (Tris/EDTA):

	To make 100 ml:
Tris, 10 mM, pH 7.4	0.5 ml of 2 M stock
0.1 mM EDTA, pH 8	20 μl of 0.5 M stock
H_2O	99.3 ml

Store at room temperature

TES buffer:

	To make 100 ml:
Tris 20 mM, pH 7.4	1.0 ml of 2 M stock
EDTA-Na_2, 0.1 mM	20 μl of 0.5 M stock
NaCl 10 mM	0.2 ml of 5 M stock

TMG buffer:

To make 100 ml:

10 mM Tris, pH 7.4	0.5 ml of 2 M stock
10 mM $MgSO_4$	1 ml of 1 M stock
0.01% gelatin	10 mg
Add H_2O to	100 ml

Tris, 2 M, pH 7.4:

To make 1 liter:

H_2O	850 ml
Tris base	242.2 g
HCl (conc.)	75 ml

Add additional concentrated HCl slowly and carefully, with stirring, until pH is 7.4 (if different pH is desired, adjust HCl volume accordingly).

H_2O to	1 liter

Triton X-100, 1%:

1 ml of Triton plus 99 ml of H_2O.

tRNA from yeast, 10 mg/ml. Used as a carrier for DNA or RNA during precipitation:

10 mg in 1 ml H_2O
Store in 1.5-ml microfuge tubes at −20°C.

Xgal solution (5-bromo-4-chloro-3-indolyl-β-D-galactoside), 10%

100 mg of Xgal in 1 ml of dimethylformamide (DMF).

Xylene cyanol, 10%:

5 g xylene cyanol. Add H_2O to make 5 ml. Resuspend before removing an aliquot.

Yeast tRNA; see tRNA, yeast.

APPENDIX II Enzymes

AMV reverse transcriptase. For formation of first-strand cDNA from mRNA template as in λgt10 and λgt11 cDNA cloning. (Seikagaku America, Inc.)

BAL 31 nuclease. A specific exodeoxyribonuclease for double-stranded species. Degrades both ends of the double strand. Requires Ca^{2+} and is inactivated by EGTA. (NEBL)

DNA polymerase I, from *E. coli.* In nick translation, works at single-stranded nicks. Also called *Kornberg enzyme.* (BM)

DNA polymerase I, large fragment (Klenow). Made from protease digestion of *E. coli* DNA polymerase I. Without 5′ to 3′ exonuclease activity. For end labeling, M13 sequencing, and probe generation. (NEBL BRL, IBI, BM)

Lysozyme. Made from egg white. Used in plasmid preparation. (P-L, Sigma)

Phosphatase. Lyophilized calf intestinal (LCIP). For removal of 5′ phosphate residues from DNA. (BM)

Pronase. A nonspecific protease. (Calbiochem)

Proteinase K. For isolation of high molecular weight DNA from cells. A nonspecific serine protease that is free of nuclease activity. (BM, Beckman)

RNase A. An endonuclease specific for RNA with no DNase activity. Store at −20°C. (BM)

RNase H. For degrading the RNA portion of DNA–RNA hybrids. From *E. coli.* (BM)

S_1 nuclease. From *Aspergillus oryzae.* Degrades single-stranded regions of DNA–RNA hybrids. Will not affect double-stranded DNA or RNA–DNA hybrids. This is a metalloenzyme used with zinc. (BM)

SP6 T3, T7 polymerase. A DNA-dependent RNA polymerase with high specificity for specific promoter sequences. Used to synthesize RNA transcripts of DNA clones. Store at −20°C. (Promega, Vector Cloning Labs, NEN)

T4 polynucleotide kinase. For 5′ end labeling of synthetic probes and DNA fragments. Store at −20°C. (BRL, P-L)

T4 DNA polymerase. For blunting recessed or protruding ends of DNA restriction fragments. Store at −20°C (BRL, IBI)

T4 DNA ligase. To join DNA fragments. Store at −20°C. (NEBL, BRL, IBI)

Terminal deoxynucleotidyl transferase. Used to add nucleotides to the 3′ end of DNA. (BRL, NEN, P-L)

Restriction endonucleases. Available from many sources, such as NEBL, BRL, IBI, BM, and P-L. Common examples include:

*Bam*HI, *Eco*RI, *Hin*cII, *Hin*dIII, *Hin*fI, *Mbo*I, *Sma*I

Hundreds of REs are commercially available. These are the ones that are used in this text. Descriptions of specific RE actions are available from the suppliers and in Section 5-4.

APPENDIX III

Suppliers of Reagents and Equipment

The following list is a compilation of the major suppliers of reagents and equipment for use in the methods contained in this book. This is not an exhaustive list or even a set of recommendations, but an indication of suppliers used by our laboratories, and is a starting point for ordering.

ELECTROPHORESIS APPARATUS

Bethesda Research Laboratories (BRL) —Assorted equipment and supplies
Life Technologies, Inc.
P.O. Box 6009
Gaithersburg, MD 20877

Bio-Rad Laboratories, Inc. —Assorted equipment and supplies
2200 Wright Avenue
Richmond, CA 94804

Hoefer Scientific Instruments —Equipment
654 Minnesota Street
P.O. Box 77387
San Francisco, CA 94107

International Biotechnologies, Inc. (IBI) —Assorted equipment and supplies
P.O. Box 1565
New Haven, CT 06506

ISCO —Power supply used in electroporation
P.O. Box 5347
Lincoln, NE 68505

LKB Instruments, Inc. —Equipment
9319 Gaither Road
Gaithersburg, MD 20877

Pharmacia, Inc. (P-L Biochemicals) (P-L) —Equipment and supplies
800 Centennial Avenue
Piscataway, NJ 08854

ENZYMES AND RELATED REAGENTS

Amersham, Incorporated —Reagents and radiolabeled chemicals
2636 South Clearbrook Drive
Arlington Heights, IL 60005

Beckman, Inc. —Enzymes and reagents, antibodies
2500 Harbor Boulevard
Fullerton, CA 92634

Bethesda Research Laboratories (See above.)

Bio-Rad (See above.)

Boehringer Mannheim Biochemicals (BM) —Assorted reagents
7941 Castleway Drive
Indianapolis, IN 46250

Calbiochem —Enzymes and chemical reagents
Division of American Hoechst
P.O. Box 12087
San Diego, CA 92112

Cooper Biomedical —Enzymes and antibodies
1 Technology Court
Malvern, PA 19355

Du Pont NEN Research Products (NEN) —Vectors, kits, radiolabeled reagents, autoradiography
549 Albany Street
Boston, MA 02118

New England BioLabs (NEBL) —Cloning vectors, enzymes, reagents
32 Tozer Road
Beverly, MA 01915

Pharmacia/P-L (See above.)

Seikagaku America, Inc. —Enzymes
9049 Fourth Street North
P.O. Box 21517
St. Petersburg, FL 33724

United States Biochemical Corporation (USB) —Enzymes and reagents
P.O. Box 22400
Cleveland, OH 44122

Vector Cloning Systems —Cloning vectors, packaging extracts and, bacterial strains
3770 Tansy Street
San Diego, CA 92121

CELLS, PLASMIDS, AND VECTORS

American Type Culture Collection (ATCC) —Cultured cells of many types
12301 Parklawn Drive
Rockville, MD 20852

Bethesda Research Laboratories (See above.)

Bio-Rad (See above.)

Boehringer Mannheim Biochemicals (See above.)

E. Coli Genetic Stock Center —Stock cells
Department of Microbiology
Yale University School of Medicine
310 Cedar Street
New Haven, CT 06510

International Biotechnology, Inc. (See above.)

New England BioLabs (NEBL) (See above.)

Pharmacia/P-L (See above.)

Promega Biotec —Cloning vectors, packaging extracts
2800 Fish Hatchery Road
Madison, WI 53711

Vector Cloning Systems (See above.)

OTHER CHEMICALS*

Aldrich —Laboratory chemicals
P.O. Box 2060
Milwaukee, WI 53201

Alfa Products —Chemicals, including methyl mercury hydroxide
152 Andover Street
Danvers, MA 01923

Collaborative Research —Peptides and nucleotide-related products
128 Spring Street
Lexington, MA 02173

Difco Laboratories —Bacterial growth chemicals
P.O. Box 1058A
Detroit, MI 48232

Fisher Scientific —Laboratory chemicals, filters, equipment
711 Forbes Avenue
Pittsburgh, PA 15219

---

* See also many of the suppliers listed above.

Eastman Kodak Company —X-ray film, developers, chemicals
343 State Street
Rochester, NY 14650

Mallinckrodt, Inc. —Chemicals and reagents
Box 5439
St. Louis, MO 63147

FMC BioProducts —Agarose
5 Maple Street
Rockland, ME 04841

Schwarz Mann —Laboratory chemicals
56 Rogers Street
Cambridge, MA 02142

Sigma Chemical Company —Laboratory chemicals
P.O. Box 14508
St. Louis, MO 63178

SPECIALTY PRODUCTS

Beckman —Centrifuges
Scientific Instrument Division
Campus Drive at Jamboree Boulevard
Irvine, CA 92713

Charles River Breeding Labs —Mice
251 Ballardvale Street
Wilmington, MA 01887

Costar —Plasticware
205 Broadway
Cambridge, MA 02139

E. I. du Pont de Nemours (Sorvall) —Centrifuges
Clinical and Instrumental Division
Bioresearch Systems
Biomedical Products Dept.
Wilmington, DE 19898

Eagle Ceramics —Glass powder fines
12266 Wilkins Avenue
Rockville, MD 20852

Falcon Labware —Cell culture plasticware
Becton Dickenson Labware
1950 Williams Drive
Oxnard, CA 93030

GIBCO Laboratories —Cell growth media and sera
Life Technologies, Inc.
Chagrin Falls, OH 44022

Fredrick Haer and Company —Capillary tubing
Brunswick, ME 04011

Kapak Corp. —Heat-sealable plastic bags
9809 Logan Ave.
Bloomington, MN 55431

KC Biologicals —Cell growth media
P.O. Box 14848
Lenexa, KS 66215

Peninsula Laboratories, Inc. —Peptides
611 Taylor Way
Belmont, CA 94002

Sarstedt —Plastic tubes
P.O. Box 4090
Princeton, NJ 08540

Schleicher and Schuell, Inc. (S&S) —Filters
10 Optical Avenue
Keene, NH 03431

Spectrum Medical Industries —Dialysis tubing
60916 Terminal Annex
Los Angeles, CA 90054

VWR Scientific —General laboratory equipment and supplies
International Headquarters
P.O. Box 7900
San Francisco, CA 94120

Whatman, Inc. —Paper filters
9 Bridewell Place
Clifton, NJ 07014

Index

T